U0323238

中国建筑装饰

CHINESE ARCHITECTURAL DECORATION

“符号中国”编写组◎编著

中央民族大学出版社
China Minzu University Press

图书在版编目(CIP)数据

中国建筑装饰：汉文、英文 / “符号中国”编写组编著. — 北京：中央民族大学出版社, 2024.8

（符号中国）

ISBN 978-7-5660-2289-9

Ⅰ. ①中… Ⅱ. ①符… Ⅲ. ①古建筑—建筑装饰—介绍—中国—汉、英 Ⅳ. ①TU-092.2

中国国家版本馆CIP数据核字（2023）第255804号

符号中国：中国建筑装饰 CHINESE ARCHITECTURAL DECORATION

编　　著　“符号中国”编写组
策划编辑　沙　平
责任编辑　陈　琳
英文指导　李瑞清
英文编辑　邱　械
美术编辑　曹　娜　郑亚超　洪　涛
出版发行　中央民族大学出版社
　　　　　北京市海淀区中关村南大街27号　　邮编：100081
　　　　　电话：（010）68472815（发行部）　传真：（010）68933757（发行部）
　　　　　　　　（010）68932218（总编室）　　　　（010）68932447（办公室）
经 销 者　全国各地新华书店
印 刷 厂　北京兴星伟业印刷有限公司
开　　本　787 mm×1092 mm　1/16　印张：12
字　　数　166千字
版　　次　2024年8月第1版　2024年8月第1次印刷
书　　号　ISBN 978-7-5660-**2289-9**
定　　价　58.00元

“符号中国”丛书编委会

唐兰东　巴哈提　杨国华　孟靖朝　赵秀琴

本册编写者

金　夏

前 言 Preface

建筑是人类实践活动的重要产物之一，更是人类文明的产物。在世界上，以宫廷木结构建筑为代表的中国传统建筑、以比萨斜塔为代表的意大利罗马建筑，以及以巴黎圣母院为代表的法国哥特式建筑并称为三大传统建筑体系，在全球范围内有着广泛的影响。中国传统建筑装饰是中国传统建筑必不可少的组成部分。这些装饰本身既有一定的实用价值，又有很强的艺术感染力，表现出了强烈的民族风格和时代特征。

随着建筑材料和技术的改进，中国传统建筑装饰产生了很多不同的变化，而这些变化又与各个历史时期的政治、经济、文化、审

Building is one of the important products of activities of human beings and furthermore, a product of human civilization. Chinese traditional architecture represented by the royal wood frame buildings, Roman architecture represented by the Leaning Tower of Pisa and French Gothic architecture represented by Notre Dame de Paris are known as the three traditional building systems that have had an extensive impact throughout the world. The traditional Chinese architectural decoration is an essential part of traditional Chinese buildings. The decoration itself has some practical value and a strong artistic appeal, showing a strong national style and characteristics of the times.

With the improvement of building materials and technology, traditional Chinese architectural decoration has changed a lot. The change is closely related to political, economic, cultural, aesthetic and other ideological factors at various historical

美等意识形态密切相关。建筑的形态、位置、等级、材料等不同，建筑装饰的形式也有所不同。本书通过对中国传统建筑装饰的历史、种类、特点、工艺材料、装饰构件等内容的梳理，系统、全面地向读者展现中国传统建筑装饰文化。希望通过本书的内容，读者更深刻地领略到中国传统建筑装饰的魅力。

periods. Forms of architectural decoration differ according to different building shapes, locations, grades, materials, etc. This book analyzes elements such as the history, kinds, features, craft materials, decorative elements of traditional Chinese architectural decoration to demonstrate to readers its culture in a systematic and comprehensive way. Hopefully readers can learn more about the charm of traditional Chinese architectural decoration through reading this book.

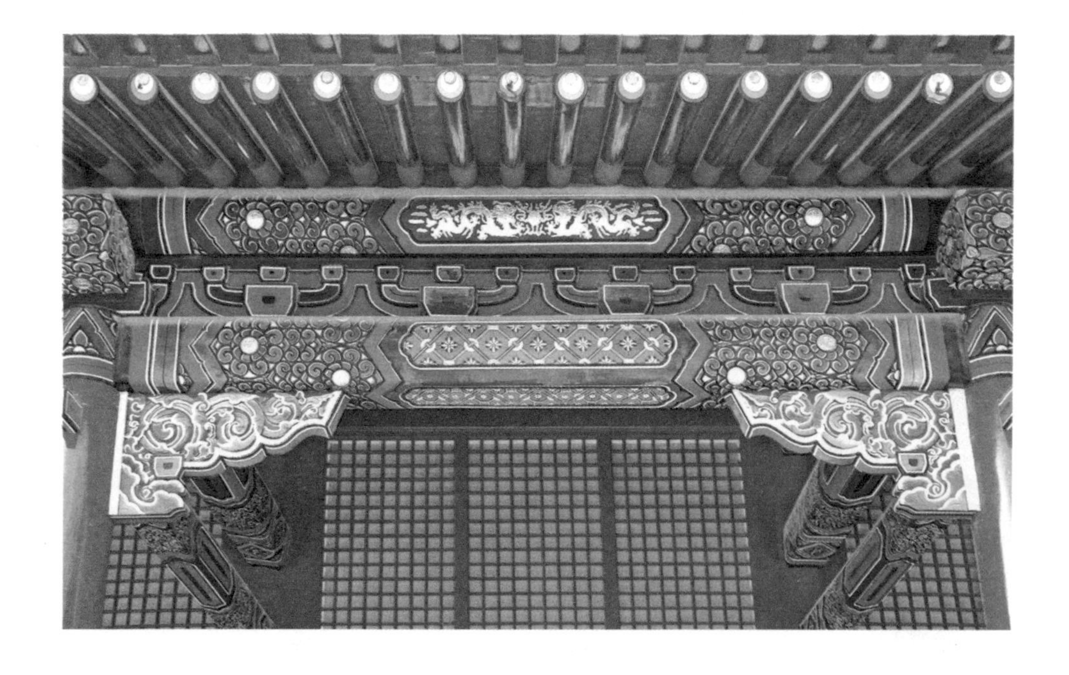

目 录 Contents

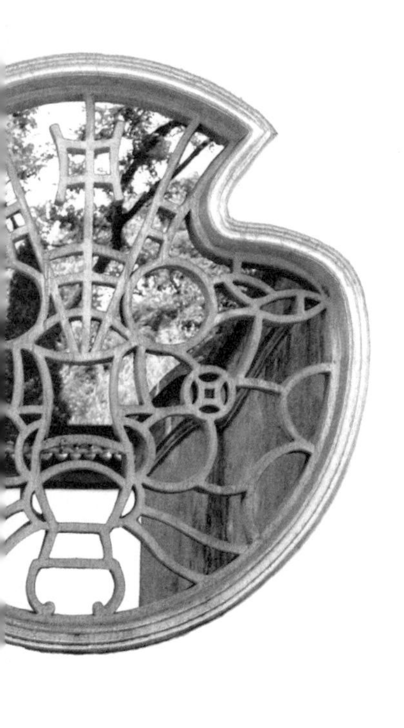

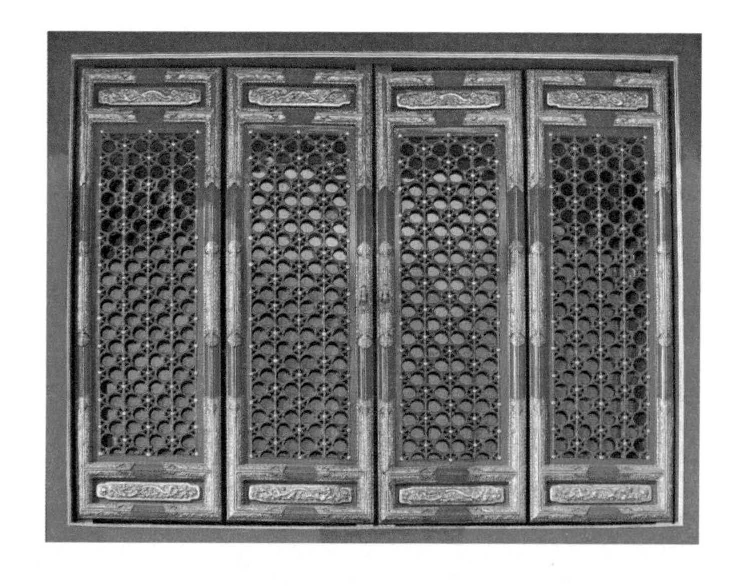

传统建筑室内装饰

传统建筑装饰源远流长的历史

Time-honored History of Traditional Architectural Decoration

中国传统建筑指从先秦到19世纪中叶以前的建筑，体现为一个独立形成的建筑体系。与世界上其他地区的人们一样，中国古代的先民也经历了从穴居、巢居，到在地面上建造房屋的过程，并形成了以木结构为主的建筑方式，进而形成世界上延续时间最长、分布地域最广、风格极其鲜明的建筑体系。建筑的装饰是伴随着建筑的出现而产生的，中国古代劳动人民凭借自己的智慧创造出了各式各样的建筑，种类繁多、五彩缤纷的建筑装饰也随之出现。这些建筑装饰起到了调整建筑构造、比例的作用，其保护建筑本身、延长建筑寿命的功用也不可忽略。

Traditional Chinese architecture refers to the building system independently-formed from the pre-Qin period to the mid-19th century. Like other peoples of the world, the ancient Chinese also went through the process of living in caves, on trees, and finally building houses on the ground and forming a building style featuring mainly wooden structures that became the most widely distributed building system with a very distinctive style which lasts the longest in the world. Architectural decoration comes into being along with the emergence of architecture. The working people of ancient China created a wide variety of architectural styles with their skill and knowledge, resulting in a wide range of colorful architectural decoration. The architectural decoration also played a role in adjusting the structure and proportion of the buildings. Its role in protecting the building and extending its service life can also not be ignored.

> 传统建筑装饰的发展

在中国古代文字甲骨文中，有不少关于建筑的文字。根据这些字形推测，公元前1000多年的商朝，人们居住的房屋有的建在台基之上，有的是干栏式住宅，虽然墙体很少装饰，但部分房屋的屋脊已经开始使用高耸的装饰构件。

大约在西周初期，除了用于覆盖屋顶的板瓦和半圆筒形的筒瓦，

• 半瓦当
Semi-circle Eaves Tile

> Development of Traditional Architectural Decoration

The ancient Chinese oracle bone inscriptions have a lot of characters related to buildings. Based on these fonts, it can be presumed that in the Shang Dynasty (1600 B.C.-1046 B.C.), some of the houses that people lived in were built on top of a podium, and some were built to be stilt houses. Although there was little decoration on the wall, tall decorative components were already used on the ridges of some houses.

In the early period of the Western Zhou Dynasty, in addition to the roof-covering plain tiles and the semi-cylindrical tiles, there emerged simple pattern tiles to decorate the cornice and protect the eaves.

In the Spring and Autumn Period (770 B.C.-476 B.C.), people began

• 鹿纹瓦当
Deer Pattern Eaves Tile

还出现了用于装饰檐口、保护屋檐的半圆形的素纹瓦当。

春秋时期，人们开始重视对居室内的装饰，出现了木建筑彩画。大约在春秋战国时期，圆形瓦当开始出现，这些瓦当有了很强的装饰性，有各种精美的纹样，如夔龙纹、夔凤纹、鹿纹等。

东汉时期到三国时期的建筑开始重视利用屋顶的形式和瓦进行装饰。屋顶形式以悬山式和庑殿式最为常见。门上多装饰有门簪，门扇多装饰以辅首。建筑上的窗常见的是直棂窗，也有的窗上装饰有其他

to focus on interior decoration and decorative color paintings for wooden buildings appeared. Round eaves tiles began to appear in around the Spring and Autumn Period to the Warring States Period (770 B.C.-221 B.C.). They were highly decorative and have various exquisite patterns, such as *Kui* loong, *Kui* phoenix and deer.

From the Eastern Han Dynasty (25-220) to the Three Kingdoms Period (220-280), people began to pay attention to the usage of roof forms and tiles for decoration. The commonly-seen roof forms are overhanging gable roof and hipped roof. Doors were often decorated with cylinders and door leaves were often decorated with door knockers. Mullioned windows were often used and sometimes other patterns were decorated on the windows. Animal prints were seen as decoration on some roof ridges.

During the Wei, Jin, Southern and Northern dynasties(220-589), as Buddhism and Taoism had more influence in the private sector, relevant patterns began to be applied to architectural decoration, such as the lotus and the Eight Diagrams patterns. Royal families often had hipped roofs. The roof ridge was often decorated with an owl-

花纹。有的屋脊开始以动物图案进行装饰。

魏晋南北朝时期，随着佛教、道教在民间的影响逐渐加大，一些与之相关的纹样开始被应用到建筑装饰中，如莲花纹、八卦纹等。贵族住宅往往使用庑殿式屋顶，屋脊上多饰以鸱尾，房屋的墙壁上多设有直棂窗。

唐代是中国建筑木结构发展的成熟时期，建筑结构与艺术达到了完美统一：斗拱硕大，使得屋檐显得十分深远；柱子下粗上细，符合

tail-shaped ornament and walls often equipped with mullioned windows.

The Tang Dynasty (618-907) witnessed the mature period of the development of Chinese wooden buildings. Architectural structure is perfectly integrated with art: the extra large brackets enabled the roof to look very far-reaching; pillars were thick at the bottom and thin at the top, which was in line with the aesthetic standards of the Tang Dynasty. In the Song Dynasty (960-1279), bricks began to be widely used during construction. Between the pillars of a residence emerged brick and wood structures and more attention was paid to the decoration than in the Tang Dynasty. The Northern Song Dynasty (960-1127) once had a provision for architectural decoration, saying that the average person may not use brackets or caissons to decorate and shall not decorate girders

• 唐代佛教建筑：陕西西安荐福寺小雁塔
Buddhist Architecture of the Tang Dynasty (618-907): Little Wild Goose Pagoda of Jianfu Temple in Xi'an, Shaanxi Province

• 宋代建筑：墨戏堂
Ink and Opera Hall of the Song Dynasty (960-1279)

唐代人以丰腴为美的审美标准。到了宋代，建筑中开始大量使用砖，住宅的柱间开始以砖木为结构，而且比唐代更加注重建筑的装饰。对于建筑装饰，北宋时曾有规定，除了宫殿、官员住宅，以及寺庙、道观，一般人的房屋不得使用斗拱、藻井，不得用彩绘装饰梁枋。

辽代和金代的建筑承袭了唐代的风格，保留了不少唐代建筑的特点，同时又受宋代建筑的影响，重视对柱间细节的装饰。元代的建筑装

with colorful paint except palaces, official residences, temples and Taoist temples.

Architecture during the Liao Dynasty (907-1125) and the Jin Dynasty (1115-1234) inherited the architectural style of the Tang Dynasty and retained some characteristics of Tang architecture. At the same time, influenced by the Song architecture, great attention was paid to detail decoration between columns. Architectural decoration had little development during the Yuan Dynasty

饰没有太大的发展，大部分建筑比较粗糙。

明代和清代，一般住宅门内设有影壁，大门、屋脊等处多有雕饰及彩绘。地面铺方砖，室内以罩、隔扇等分隔空间。紫禁城是这一时期宫殿建筑达到顶峰的代表性作品，传统建筑的装饰也被发挥到了极致。

(1206-1368) and most of the styles were relatively primitive.

During the Ming Dynasty (1368-1644) and the Qing Dynasty (1616-1911), there was usually a screen wall inside a residential door and carved ornamentation and color paintings on the door and the roof ridge. Square bricks were used to pave the ground and covers and partition boards were used to divide up interior space. The Forbidden City is a masterpiece exemplifying the climax of the development of palatial architecture during the period and the traditional architectural decoration was also brought into full play.

紫禁城

紫禁城是中国明、清两个朝代的皇宫。明朝的第三位皇帝朱棣在夺取帝位后，决定迁都北京，并开始营造宫殿，至明永乐十八年（1420年）落成。有这样一种说法，依照中国古代星象学说，紫微垣（即北极星）位于中天，乃是天帝所居之处，天人对应，因此皇帝的居所又称“紫禁城”。

紫禁城位于北京，现称“故宫”。紫禁城南北长961米，东西宽753米，占地面积达72万多平方米，有房屋9000多间。四面环有城墙和护城河，城墙四边各有一门，南为午门，北为神武门，东为东华门，西为西华门。紫禁城的南半部以太和殿、中和殿、保和殿三大殿为主，两侧为文华殿、武英殿，是皇帝举行朝会的地方，称为“前朝”。北半部则以乾清宫、交泰殿、坤宁宫三宫和御花园为中心，两旁有东西六宫，其外东侧有奉先、皇极等殿，西侧有养心殿、雨花阁、慈宁宫等，是皇帝和后妃们居住、举行祭祀和宗教活动，以及处理日常政务的地方，称为“内廷”。

紫禁城的宫殿都是木结构、黄琉璃瓦顶、青白石底座，饰以绚丽夺目的彩画，整体建筑风格雄伟、富丽堂皇。

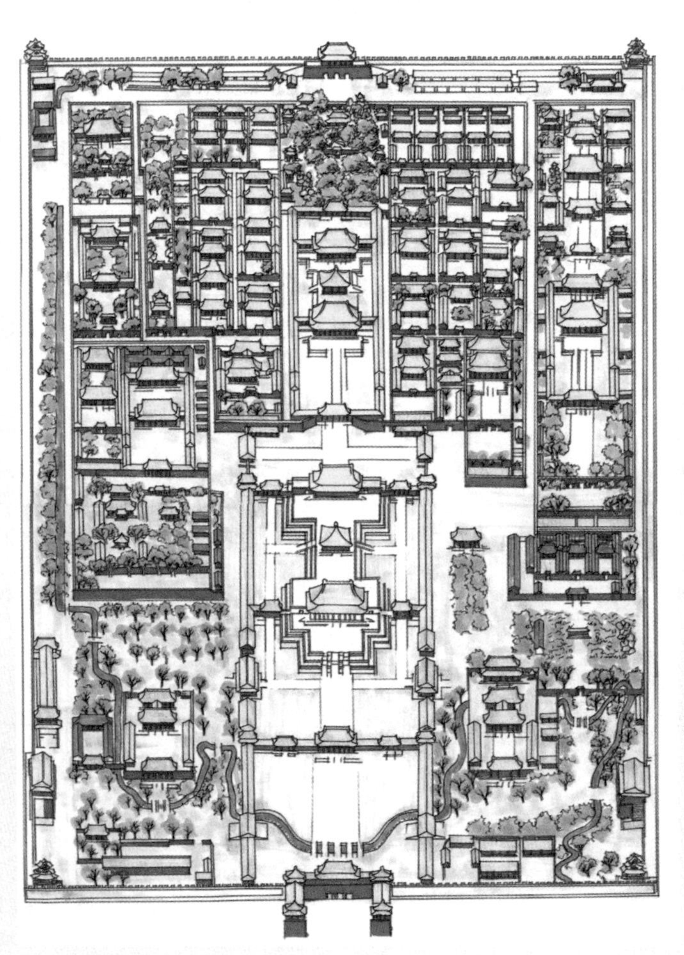

• 紫禁城示意图

Sketch Map of the Forbidden City

The Forbidden City

The Forbidden City was the Imperial Palace during the Ming Dynasty (1368-1644) and the Qing Dynasty (1616-1911). It was completed in the 18th year (1420) during the reign of the third Ming-dynasty emperor Yongle who upon usurping the throne, had decided to move the capital northward to Beijing. It is said that, according to ancient Chinese astrology, the Purple Forbidden Enclosure (i.e. the North Star) is located in the meridian passage where the Supreme God lives. Therefore, the mortal equivalent of the Supreme God, the emperor, should live in a residence called "Forbidden City".

The Forbidden City is located in China's capital Beijing, and was hence called the Imperial Palace. It is 961 meters long from north to south and 753 meters wide from east to west, covering an area of more than 720,000 square meters. It has more than 9000 houses. It is surrounded by a wide moat and a city wall with a gate on every side. At the southern end is the Meridian Gate. To the north is the Gate of Divine Prowess. The east and west gates are called the East Prosperity Gate and West Prosperity Gate. The Hall of Supreme Harmony, the Hall of Central Harmony, and the Hall of Preserving Harmony are located on the central axis of the southern end of the Forbidden City. Lying on the two sides of the central axis are the Hall of Literary Brilliance and the Hall of Martial Valor. This is called the "Front Court" and used for ceremonial purposes. The northern end of the Forbidden City is called the "Inner Court" where the Palace of Heavenly Purity, the Hall of Union, the Palace of Earthly Tranquility and the imperial garden lie on its central axis. It also includes the Hall for Ancestral Worship and the Hall of Imperial Supremacy on the east and the Palace of Mental Cultivation, Tower of Raining Flowers and Palace of Compassion and Tranquility on the west, making up altogether six palaces. The Inner Court is the residence of the Emperor, the Queen and the imperial concubines, and was used for holding ceremonies to worship heaven, conducting political activities and handling day-to-day government affairs.

The Forbidden City is of wood structure and features yellow glazed tile roofs, and a green whitehead base decorated with magnificent color paintings. The architectural style is majestic and splendid.

> 传统建筑装饰的特点

无论是官式建筑还是民间建筑，它们的风格特点在很大程度上来源于装饰。装饰造就了中国传统建筑富有特征的外观，让建筑更具有思想内涵和艺术性。从形态上划分，中国传统建筑可分为城池、皇家建筑、礼制坛庙、园林、民居、陵墓、寺庙、道观、塔、牌坊、桥梁等。这些类别的建筑大多结构奇巧，装饰精美，形成了自己独特的形态和风格：以木材、砖瓦为主要建筑材料；采用平面布局体现出一种简明的组织规律；建筑造型优美，装饰丰富多彩，同时特别重视与周围自然环境的协调。

色彩特点

中国人在世界上最早懂得使用

> Characteristics of Traditional Architectural Decoration

Whether it is the style of an official building or a civil construction, the characteristics are largely derived from decoration. Decoration gives birth to the characteristic appearance of traditional Chinese architectures and instills more ideological content and artistic nature into buildings. Judging from the forms, traditional Chinese architecture can be divided into city walls, royal buildings, ritual temples, gardens, civic houses, tombs, Buddhist temples, Taoist temples, towers, memorial archways, bridges, etc. Most of these architectural categories have exquisite structures and delicate decorations and have formed their own unique shapes and styles: wood and brick are used as the main construction materials, the plane layout is simply organized, the shape is elegant and

色彩，色彩文化是中国传统文化的重要组成部分。古人很早就确立了色彩结构，以青、赤、黄、白、黑五种颜色为正色，并把中国人关于自然、伦理、哲学的多种观念融入色彩中。中国传统建筑以土木结构为主，建筑所使用的色彩极大地丰富了建筑本身，并且具有深刻的文化内涵。金黄的琉璃瓦、朱红的门窗、白墙和黛瓦，以及五彩缤纷的彩画额枋，让中国传统建筑焕发出了无限的活力。

在古代，人们把较浅的蓝色称为“青色”，代表着东方方位。建筑装饰常用的蓝色是群青，也叫“云青”，这是一种色泽鲜艳的蓝色。屋顶上的吻兽、瓦当也常被着蓝色，而且在建筑彩画中，蓝色经常与青莲等色搭配使用，具有很好的装饰性。

红色可以说是中国的代表色，象征着吉祥、喜庆、勇敢、正义。古代的中国人，从洞房花烛到金榜题名，从衣着到住所，对红色的崇尚随处可以体现。皇宫建筑的宫墙、门窗、柱子、额枋及民居的大门大都被漆成红色，而在斗拱、天花板等建筑构件的彩绘中，红色的

the decoration is colorful. At the same time, special attention was paid to the coordination of the buildings with their surrounding natural environment.

Characteristics of Colors

The Chinese is one of the first in the world to know how to use colors, and color culture has become an important part of Chinese traditional culture. From time immemorial, ancient Chinese people established a color structure that treats blue, red, yellow, white and black as pure colors and correlates them with the five elements of metal, wood, water, fire and earth. Various concepts of Chinese people about the natural universe, ethics and philosophy were integrated into colors. The main structure of traditional Chinese architecture is civil structure. The colors used on the architecture have greatly enriched the building itself and have profound cultural connotations. Golden glazed tiles, vermilion doors and windows, white walls, black tiles and architraves with color paintings enable the traditional Chinese architectures to shine with vigor and vitality.

In ancient times, blue was commonly referred to as "cyan" which represents the position of the East. The blue

• 北京北海公园陟山桥坊彩绘
Color Painting of Memorial Gate in Beihai Park, Beijing

使用频率也很高。

在古代中国，黄色象征着权力、富贵、光明和智慧。黄色曾是皇族的专用色，还代表着孕育万物的土地。

在传统建筑装饰当中，白色的使用频率不是很高，多用于墙壁，例如中国南方徽州民居的显著特征之一就是白色的墙壁。

在远古时期，中国人崇尚黑色，以黑为贵，认为它在方位上象征着北方。民居、园林等建筑的屋顶常用黑色的瓦件即“黛瓦”（黛是一种青黑色的矿物颜料）进行装

color commonly used in architectural decoration is ultramarine, also called "*Yunqing*" which is a brightly-colored blue. The loong-head ornament and eaves tiles on the roof ridge were often painted in blue. Blue and pale purple were usually used together in the color paintings of buildings and have good decorative effect.

Red is almost the representative color of China. It symbolizes good luck, happiness, bravery and justice. The habit of loving red by ancient Chinese people can be seen everywhere, be it on a wedding night, after one's success in a government examination and from

• 乾清门檐下被漆成大红色，象征皇权的威严
Eaves of the Gate of Heavenly Purity is Painted Red, Symbolizing Imperial Majesty

饰。在彩画当中，黑色的墨线则起到了很好的色彩过渡作用。

青绿色常被用于寺庙等建筑物的彩画当中，建筑装饰当中常见的表现形式还有竹青绿的琉璃瓦当、琉璃螭兽（中国古代神话传说当中一种没有角的龙）等。

图案特点

传统建筑装饰中的吉祥图案体现了中国传统文化的精髓，是中国几千年历史文化的缩影。从这些吉

wearing clothes to decorating residences. Most of the walls, doors, windows, pillars and architraves of an imperial palace and the residential doors are painted red. In addition, red is frequently used in painting brackets, ceilings and other architectural elements.

In ancient China, yellow symbolizes power, wealth, light and wisdom. Yellow used to be exclusive to the royal family and represents the land that nurishes all things.

White is not frequently used in traditional architectural decoration except

• 北京故宫奉先殿的琉璃瓦顶
The Hall for Ancestry Worship in the Forbidden City in Beijing Covered on Top by Glazed Tiles

• 上海玉佛寺的黄色墙壁
Yellow Wall of the Jade Buddha Temple, Shanghai

带有白色墙壁的徽州古村落
Ancient Village with White Walls in Huizhou

祥图案中，我们不仅能看到各个时期的建筑特色，还能看出民俗及外来文化对中国古典建筑的影响。

吉祥图案被用于建筑装饰的历史，可以追溯到商周时期。那时的人们在瓦当上刻画各种吉祥图案，以求建筑免于灾祸。秦汉时期是中国传统建筑发展史上的第一个高潮，帝王大兴土木，在建筑装饰上使用的吉祥图案也逐渐增多。隋唐

for walls. For example, one of the salient features of residences in Huizhou, in southern Chinese, is white walls.

In ancient times, Chinese people advocated and valued black, and hold the belief that it is a symbol of the North. The roofs of houses and gardens usually use black tiles called "*Daiwa* (*Dai* is a bluish black mineral pigment)" for decoration. In color paintings, the black ink line played an important role in realizing

黑瓦铺顶

Black Tiles Are Paved on the Roof Top

时，建筑装饰用的吉祥图案已形成了独特的风格，并影响到了周边其他国家。宋元时期的建筑更加注重吉祥图案的应用，呈现出活泼、明丽的风格。明、清两代，吉祥图案

绿色琉璃瓦当

Green Glazed Tiles

color transition.

Green is often used in color paintings in temples and other buildings. Bamboo green glazed tiles and a glazed *Chi* (a hornless loong in ancient Chinese myths and legends) are also commonly seen in architectural decoration.

Characteristics of Patterns

The auspicious patterns in traditional architectural decoration are a reflection of the essence of traditional Chinese culture and a miniature of thousands of years of Chinese history and culture. From these auspicious patterns, one can not only see the architectural feature of each era, but also the influence of folk

题材丰富，使用范围广泛，花草、树木、飞禽、走兽甚至器物都被纳入吉祥图案。建筑装饰中的吉祥图案既融中国绘画、书法、工艺美术于一身，表达了古人对美的认知和感悟，又具有极高的观赏价值，从而成为中国传统建筑装饰的显著特点。

按照题材的不同，建筑装饰图案可分为祥禽、瑞兽、花木、器物、图文、人物等，“纹必有意，意必吉祥”，造就了各式各样内容丰富的传统纹饰。

祥禽、瑞兽

在传统建筑装饰当中，祥禽、瑞兽占据着重要的地位，常被用于石雕、木雕、陶塑等。祥禽、瑞兽包括龙、凤、狮子、虎、麒麟、龟等，是人们祈盼平安和幸福的象征，备受喜爱和推崇。

customs and foreign cultures on classical Chinese architectures.

It can be dated back to the Shang Dynasty (1600 B.C.-1046 B.C.) and the Zhou Dynasty (1046 B.C.-256 B.C.) when buildings were decorated with auspicious patterns. People then depicted various auspicious patterns on eaves tiles in order to protect the buildings from disaster. The Qin Dynasty (221 B.C.-206 B.C.) and the Han Dynasty (206 B.C.-220 A.D.) witnessed the first climax in the developement of traditional Chinese architecture. Emperors launched large-scale construction projects and architectural decoration gradually adopted more auspicious patterns. During the Sui (581-618) Dynasty and the Tang Dynasty (618-907), auspicious patterns used for architectural decoration formed a unique style and influenced other neighboring countries. The architecture of the Song Dynasty (960-1279) and the Yuan Dynasty (1206-1368) paid more attention to the application of auspicious patterns, demonstrating a bright and lively style. The subject matters of the auspicious patterns were enriched and used on a large scale during the Ming Dynasty

• 天安门前的华表

天安门前的华表既是装饰物，又是皇家建筑群的标志。柱上雕有一条龙，盘旋而上，龙鳞清晰可见，威武而庄严。龙是中国神话传说中的一种神异动物，体长而威猛，能腾云、驾雾，会兴风、降雨。龙被分为四种，即身上带鳞的蛟龙、有翅膀的应龙、头上无角的螭龙和头上有角的虬龙。中国传统建筑中常见的龙形图案有团龙、蟠龙、坐龙、夔龙、双龙戏珠、鸱尾、云龙纹、草龙纹、拐子龙等。在中国古代，龙一般为皇帝所专用，象征着皇家的权威、高贵；但是在民间，龙常用于表达吉祥和喜庆，常用于寺庙、祠堂等建筑上。

Cloud Pillar before Tian'anmen

The cloud pillar before the Tian'anmen is both an ornament and a symbol of a royal architectural complex. The column is carved with a loong coiling its body up. The clear loong scales make it mighty and solemn. The loong is a miraculous animal in Chinese myth and legend. It is long and mighty and can mount the clouds, ride the mist, arouse the wind and cause the rainfall. The loong can be classified into four categories, namely, scaled loong (*Jiao* loong), winged loong (*Ying* loong), hornless loong (*Chi* loong) and horned flying loong (*Qiu* loong). The commonly seen Loong-shaped patterns in traditional Chinese architecture includes the coiled loong, the curled-up loong, the sitting loong, the *Kui* loong, two loongs playing with a pearl, owl tail, cloud and loong, the grass loong and the *Guaizi* loong. In ancient China, the loong is usually used exclusively by emperors and is a symbol of royal authority and highness. However, the loong is often used by folk to represent good luck and festive atmosphere in temples, ancestral halls and other buildings.

• **台湾天后宫屋脊上的凤**

凤凰乃百鸟之王，雄为凤，雌为凰，喻示着和平、美好。在皇家建筑中，尤其是在清代，凤凰装饰多用于后妃居所。凤还有“贤才逢盛世”的意思，常见的凤凰题材装饰主要有丹凤朝阳、凤戏牡丹、百鸟朝凤、龙凤呈祥、凤栖梧桐等。

Phoenix on the Roof of Tianhou Palace, Taiwan Province

The phoenix is the king of birds. The male phoenix is called *Feng* and the female phoenix is called *Huang*. The phoenix is a symbol of peace and beauty. In royal buildings, especially in the Qing Dynasty, phoenix patterns were used for the decoration of the residence of the Queen and the imperial concubines. The phoenix also has the connotation of talented people meeting the golden age. The commonly seen themes of the phoenix patterns are "the scarlet phoenix flies towards the sun" "the phoenix plays with peony" "all birds paying homage to the phoenix" "the loong and the phoenix bring prosperity" and "the phoenix dwells on Chinese parasol tree".

(1368-1644) and the Qing Dynasty (1616-1911). For example, flowers, trees, animals and even utensils were included in auspicious patterns. The auspicious patterns in architectural decoration integrated Chinese paintings, calligraphy, arts and crafts to express the ancient people's understanding in beauty. These patterns boast very high ornamental value and have become a salient feature of architectural decoration.

Architectural decorative patterns can be classified into auspicious animals, flowers, trees, objects, graphics and characters according to different themes. The idea that "patterns must connote meanings of luck" has injected meanings into various traditional ornamentations.

Auspicious Animals

In traditional architectural decoration, auspicious animals hold an important position and are commonly used in stone carving, wood carving and ceramics, etc. Auspicious animals including the loong, phoenix, lion, tiger, kylin and tortoise which symbolize people praying for peace and happiness, are much loved and respected.

绍兴太师少师石雕

在绍兴太师少师石雕中，狮子表情生动，造型活泼，细节十分讲究。狮子又称“瑞狮”，在宫殿、寺庙、宗祠、民居、园林中都比较常见，多被成对摆放在大门前。在古人心中，狮子是权贵的象征，喻示神圣而不可侵犯，同时又有镇宅、求得平安的寓意。常见的题材主要有狮子滚绣球、太师少师等。南北方石狮造型不同，北方狮子多威武，南方狮子多活泼。

Stone Carving Lions in Shaoxing

The stone carving lions of Shaoxing in Zhejiang Province have expressive and animated faces and are delicately designed. The lion is also known as "auspicious lion" and is commonly seen in palaces, temples, ancestral halls, residential buildings and gardens, and is usually placed in front of a gate in pairs. In the hearts of the ancient Chinese, the lion is a symbol of the rich and the powerful. It has a sacred and inviolable meaning and can also be used to keep away evil spirits and seek peace. The common themes are "lion frolicking with the festival ball", and "grand lion and little lion". Lion shapes from the south and the north are different from each other, the north being powerful and the south being active.

铜铸麒麟

麒麟为神话传说中的瑞兽，与龙、凤、龟并称为“四灵兽”，在宫殿、园林、民居等建筑中被广泛使用，在建筑贴面、砖墙、垂花门、天花、额枋等处也较为常见，有吉祥、太平、福禄等寓意。

Bronze Kylin

The kylin is an auspicious animal in Chinese myths and legends and is known as one of the "Four Sacred Beasts" together with loong, phoenix and tortoise. It is widely used in palaces, gardens and residential buildings, and commonly seen on building veneer, brick walls, festooned doors, ceilings and architraves. It means good luck, peace and longevity.

- **北京北海公园琼岛永安寺山门殿前的铜龟**

人们把龟视为长寿的象征，与龙、凤、麒麟并称为“四灵兽”。龟背纹样还常出现在建筑的砖、石等构件上，起装饰作用。

Copper Turtle in Front of the Mountain Gate Hall of Temple of Eternal Peace on the Jade Islet in Beihai Park, Beijing

The turtle is treated as a symbol of longevity and known as one of "four sacred beasts" together with the loong, phoenix and kylin. Turtle patterns often appear on the bricks, stones and other components of the building for decorative purposes.

- **唐代石虎**

虎是猛兽，古人认为虎是降伏“魔鬼”的“神兽”。早在战国时期，虎纹就被用来装饰瓦当。虎的石雕常被用于宫殿、园林、陵墓等。人们认为它既能保佑家宅吉祥、平安，又可消除灾病。民间又有“云从龙，风从虎”的说法，虎纹还有风调雨顺、国泰民安的寓意。

Stone Tiger in the Tang Dynasty (618-907)

The tiger is a beast regarded by the ancient Chinese as a divine animal subduing demons and ghosts. As early as in the Warring States period (475 B.C.-221 B.C.), the tiger pattern is used to decorate eave tiles; tiger-shaped stone carving is often used in palaces, gardens, tombs and other buildings. Tigers are believed to bless houses with good fortune and peace and eliminate the plague. In folk culture, it is said that "cloud is controlled by loong and wind is controlled by tiger". Therefore, the tiger pattern also symbolizes good weather, peace and prosperity.

• **连年有余木雕**

鱼与“余”“玉”谐音，象征着富余、财源滚滚。常见的题材主要有金玉满堂、连年有余、吉祥如意、鲤鱼跳龙门等。

Year-on-year Surplus Wood Carving

"Fish" "surplus" and "jade" are homonymic in Chinese, symbolizing surplus and constant fortune. The common themes are "great wealth" "year-on-year surplus" "good luck" and "carps jumping over the loong gate".

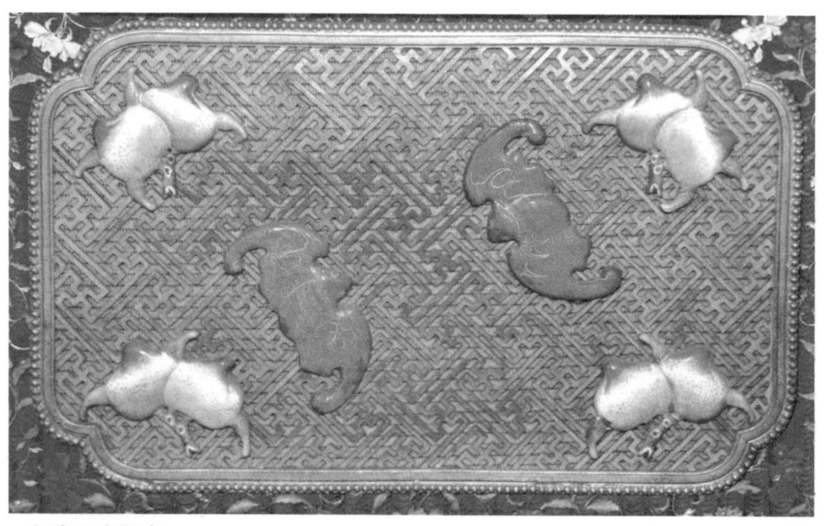

• **福寿双全图案**

蝙蝠的“蝠”与“福”同音，喻示着幸福美满、吉祥如意。建筑装饰当中的蝙蝠形象多经过美化。

Pattern of Happiness and Longevity

"Bat" and "happiness" are homonymic in Chinese, symbolizing happiness and good luck. Most of the bat images in architectural decoration have been beautified.

北京北海公园琼岛永安寺山门殿前的铜鹤

鹤是长寿的鸟类，在古代神话传说中是仅次于凤的“仙禽”，被认为是超凡脱俗的禽类。鹤又因体态高洁、优雅 而成为文人、雅士志向高远、不与凡夫俗子同流合污的品格的象征。

Copper Crane in Front of the Mountain Gate Hall of Temple of Eternal Peace on the Jade Islet in Beihai Park, Beijing

The crane is a long-lived bird and an immortal animal second only to the phoenix in ancient myths and legend. It is considered to be extraordinarily refined. Being noble and elegant, it has become a symbol of virtuous character for ambitious and undefiled literati.

花木

以花草、树木为题材的吉祥图案在古典建筑中最为常见，如被称为“百花之王”“富贵花”的牡丹，被誉为“四君子”的梅、兰、竹、菊等。它们或者借用花木本身的吉祥寓意，或者利用这些植物的特征组成了吉祥图案，较为常见的包括福寿三多、连中三元、榴开百子等。

Flowers and Trees

Auspicious patterns themed on flowers and trees are commonly seen in classical architecture. For example, the peony which is called the "King of Flowers" and the "fortune flower", and plum blossom, orchid, bamboo and chrysanthemum which are known as the "four gentlemen". Borrowing auspicious meanings of flowers and trees or using the characteristics of these plants, the commonly seen ones of which include "good fortune and longevity" "get the three highest literary degrees in succession" "pomegranate revealing 100 seeds", etc.

北京北海公园延楼外檐上的国色天香彩画

牡丹是中国特有的名贵花卉，雍容华贵，被认为是富贵的象征，又称“富贵花”。

Color Painting of National Beauty and Heavenly Fragrance on the Roof Ridge of the High Building in Beihai Park, Beijing

The peony is a rare and elegant flower unique to China. It is considered a symbol of wealth, hence its another name, "fortune flower".

苏州拙政园建筑屋脊上的榴开百子

古人视多子石榴为吉祥物，寓意为多子多福、儿孙满堂。

Pomegranate Revealing 100 Seeds on the Roof of the Humble Administrator's Garden, Suzhou

The pomegranate is a plant that produces numerous seeds and is seen as a mascot by ancient Chinese people, meaning "the more sons, the more blessings" "plenty of progeny".

- **梅兰竹菊木雕**

梅，寒冬依旧盛开，能够忍受寒冷；兰，生于深林，不因为没人赏识而停止散发芬芳；竹，弯而不折，坚韧而挺拔；菊，秋季仍旧傲霜而立，品质高洁。梅、兰、竹、菊并称为“四君子”，有洁身自好、品行高洁、坚贞不屈、虚心、正直之寓意。

The Wood Carving of Plum Blossom, Orchid, Bamboo and Chrysanthemum

The plum blossom is a plant still in full bloom in the cold winter and is able to endure it. The orchid originates in deep forests and emits a strong fragrance despite few appreciators. The bamboo can bend but not yield. The chrysanthemum is a noble and virtuous plant that stands proudly against the chilly frost in autumn. Plum blossom, orchid, bamboo and chrysanthemum are collectively called the "four gentlemen", meaning that they are virtuous, unyielding, humble and upright.

器物

用喻示吉祥的器物组成吉祥图案，或借用谐音与其他物体组合成吉祥图案是中国古典建筑中较为常见的吉祥图案形式。器物组合而成的吉祥图案中最常见的是博古图，它可以代表家族的文学修养或闲情逸致，常见于士大夫或达官贵人住

Objects

Forming auspicious patterns using auspicious objects, or combining auspicious objects with other objects with homophonic names are common forms of patterns in classical Chinese architecture. The most commonly seen auspicious pattern formed by objects is Diagram of Knowledge, which

宅的装饰。除此之外，常见的吉祥图案还包括八宝图、聚宝盆、铜钱、琴、棋、书、画等，这些吉祥图案代表着百姓对美好生活的期盼。

represents the literary accomplishment and the leisurely and carefree mood of the family. It is commonly seen in the residential decoration of scholar-officials or dignitaries. In addition, the usually seen auspicious patterns also include the Figure of Eight Treasures, the Pot of Gold, coins, Zither, chess, calligraphy and painting, etc. They symbolize people's wishes for a better life.

• **苏州拙政园双钱纹铺地**

以铜钱纹装饰建筑很早就已开始，铜钱也常与喜鹊、蝙蝠等组合成吉祥图案，最为常见的是铜钱纹花窗和铜钱纹铺地，喻示财源滚滚、衣食无忧。

Copper-coin-paved Floor in the Humble Administrator's Garden, Suzhou

The copper coin pattern has long been used to decorate buildings. Sometimes copper coins will be used to form auspicious patterns together with magpies and bats. The most commonly seen are copper coin pattern windows and floors. They mean constant fortune and comfortable living.

- **山西乔家大院主楼雀替上的木雕博古图**

  博古图案指包括瓷器、玉器、奇石、铜器等各种古器具，以及各种花卉、草木的装饰图案，常用于建筑及家具的装饰。其寓意为清新、高雅、超凡脱俗。

  The Wood Carving of Diagram of Knowledge on the Sparrow Brace of the Main Building of the Qiao's Courtyard in Shanxi Province

  The diagram of knowledge pattern refers to the decorative patterns of various ancient objects including porcelain, jade articles, strange stones and bronze utensils, and a variety of flowers and trees. They are commonly used for the decoration of architecture and furniture. The Diagram of Knowledge pattern has a connotation of elegance and refinement.

文字

以文字为主的吉祥图案的表现形式更加直接。如喻示福运当头的“福”字、健康长寿的“寿”字、喜事连连的“喜”字，以及回字纹

Characters

The character-themed auspicious patterns have a more direct form of expression. For example, the word "*Fu*" which represents happiness and good fortune, "*Shou*" symbolizes health and longevity,

等。随着佛教影响的扩大，佛教中的一些吉祥符号也融入传统建筑图案中。

"*Xi*" means the reoccurring of happy events and the pattern resembling the Chinese character "*Hui* (回)" means good luck. In addition, as the influence of Buddhism expanded, auspicious symbols of Buddhism were also integrated into traditional architectural patterns.

- **平遥古建筑门前影壁上的福字**

福字在中国传统建筑中甚为常见，无论是民间建筑还是宫廷建筑，常出现在门上或者墙壁上，喻示福运当头、福气临门。人们常常以这种方式祈祷幸福、美满、吉祥。

"*Fu*" Character on the Screen Wall in Front of the Gate of Pingyao Ancient Buildings

The "*Fu*" character is most common in traditional Chinese architecture. Be it a civil or an imperial building, the "*Fu*" character often appears on doors or walls, meaning the coming of blessings and good fortune, and expresses people's prayers for happiness and good luck.

人物

传统建筑装饰还常常以神话传说和历史人物为题材，例如嫦娥奔月、牛郎织女、八仙过海、水漫金山、钟馗捉鬼、二十四孝等，寄托着人们对幸福生活的向往，教育子孙孝敬父母、尊敬长辈等。

Figures

Traditional architectural decoration is also often themed on the characters of myths, legends and historical records, such as Chang'e Flies to the Moon, The Cowherd and the Weaving Maid, The Eight Immortals Crossing the Sea, Flooding the Golden Mountain Temple, Zhong Kui the Ghost Catcher and the 24 Filial Exemplars. It carries people's expectations of leading a happy life, educating children, supporting parents and respecting elders.

• 苏州木渎古镇秀野竹堂上的八仙

“八仙”分别指铁拐李、汉钟离、吕洞宾、张果老、曹国舅、韩湘子、蓝采和、何仙姑，是民间传说中的八位道教“神仙”。此题材表达了人们对长寿的渴望。

The Eight Immortals in the Xiuye (Beautiful Field) Room in Mudu Town, in Suzhou

The Eight Immortals refer to Tieguai Li, Han Zhongli, Lü Dongbin, Zhang Guolao, Cao Guojiu, Han Xiangzi, Lan Caihe and He Xiangu. They are eight Taoist immortals in folklore. The theme reflects people's wishes for longevity.

材料特点

由于中国传统建筑以木结构为主，辅以砖、瓦、石等材料，根据建筑材料的不同，传统建筑装饰也被用以不同的材料和方法，如石雕、砖雕、木雕、陶塑、灰塑、彩绘等。

石雕

石雕是建筑装饰中被使用得最广泛的品种。随着佛教传入中国，佛教建筑开始盛行，石窟、佛像、佛塔等都以石雕进行装饰。除了石

Characteristics of Materials

Traditional Chinese architecture mainly consists of wooden structures supplemented by various materials such as bricks, tiles and stones, etc. As a result, traditional architectural decoration adopted different materials and methods such as stone carving, brick carving, wood carving, pottery sculpture, plaster sculpture and color painting according to the building materials.

Stone Carving

Stone carving is the most widely used kind of architectural decoration. With the introduction of Buddhism into China, Buddhist architecture began to flourish. Stone carving decoration began to appear on grottoes, Buddhist statues and pagodas. In addition to perrons, platforms, Buddhist Sumeru bases, stone balustrades and decorated archways, ornamental columns, stone lions, stone tablets and stone buildings all featured stone carving decoration. Later on, a large

• 山西王家大院门墩——圆雕石狮
Door Mound of the Wang's Courtyard in Shanxi Province—Circular Carving Lion

台基、高台、须弥座、石栏杆、牌楼，还有华表、石狮、石碑、石幢等。后来，石雕被普遍用于建筑及其构件。石雕类别丰富多样，包括圆雕、浮雕、透雕等。

圆雕又称“立体雕刻”，即在各个角度都要进行雕刻。浮雕即在平面上雕刻出凹凸起伏的形象。透雕也称“镂雕”，即在浮雕的基础上，对其背景部分采取镂空的处理方式。

variety of stone carving was commonly applied to buildings and architectural components. These included circular carving, rilievo and openwork carving.

Circular carving is also known as "three-dimensional" carving, meaning that the engraving should be carried out at all angles. Rilievo refers to carving patterns of ups and downs on a flat surface. Openwork carving is also known as "through carving" which is to hollow out the background of a work piece on the basis of rilievo.

• 山西乔家大院建筑栏杆上的锦鸡牡丹浮雕

Golden Pheasant Rilievo on the Building Balustrade of the Qiao's Courtyard in Shanxi Province

苏州拙政园建筑屋脊上的丹凤朝阳石雕
Stone Carving of "the Scarlet Phoenix Flies toward the Sun" on the Roof Ridge of the Humble Administrator's Garden, Suzhou

砖雕

砖雕是模仿石雕而出现的一种雕饰类别，由于比石雕更加省工、经济，在建筑中逐渐被采用。砖雕是由东周时期的瓦当、空心砖和汉代画像砖发展而来的，形成于北宋，被用作墓室壁的装饰品。明代，砖雕由墓室砖雕发展为建筑装饰砖雕；清代则是砖雕发展的巅峰时期。

Brick Carving

Brick carving is a type of ornamentation that imitates stone carving. It was gradually adopted over stone carving in buildings due to its labor-saving and economic features. It gradually developed from eave tiles and hollow bricks of the Eastern Zhou Dynasty (770 B.C.-256 B.C.) and portrait bricks of the Han Dynasty. It took shape in the Northern Song Dynasty (960-1127) and was used as tomb wall decoration. In the Ming Dynasty, it developed from tomb brick carving to brick carving in wider architectural decoration. The Qing Dynasty was the peak period for the development of brick carving.

画像砖

画像砖起源于战国时期，盛行于两汉，多用于墓室壁画，在宫室建筑中也被使用。画像砖主要以木模压印，然后经火烧制而成；也有的表现为在砖上刻出纹饰。画像砖画面的表现形式有浅浮雕、阴刻线条和阳刻线条。多数画像砖为一砖一个画面，也有一砖有上、下两个画面的形式。题材内容多为播种、收割、舂米、酿造、盐井、宴乐、杂技、舞蹈、民间神话传说中的人物形象等。

Portrait Brick

Portrait brick originated from the Warring States Period (475 B.C.-221 B.C.) and prevailed in the Western Han Dynasty (206 B.C.-25 A.D.) and the Eastern Han Dynasty (25-220). It was mostly used for tomb murals and sometimes for imperial buildings. Portrait brick is made by being fired after embossing with wood mold. Some are carved with patterns. The forms of expression of portrait brick are bas-relievo, intaglio engraving and raised engraving. The majority of portrait bricks are engraved with only one image but some are engraved with two images. The themes are mostly about planting, harvesting, husking, brewing, salt making, feasting, acrobatics, dances, and characters in folk myths and legends.

• 东汉时期的舂米画像砖

Portrait Brick Showing Husking in the Eastern Han Dynasty (25-220)

砖雕有两种做法：一种为“硬花活”，以雕刻工具在特制的水磨青砖上直接雕刻而成；另一种叫“软花活”，纹样以纸浆灰堆塑而成，制作方法不是雕刻，而是塑。砖雕一般用于门楼、影壁、祠堂、戏台、山墙、寺庙、牌坊等建筑的装饰，雕刻精巧，题材丰富。

There are two ways of making portrait bricks. One is "hard decorations" that are carved directly on specially-made blue bath brick using carving tools. The other is "soft decorations" whose patterns are made through molding pulps instead of carving. Brick carving is usually used on gate towers, screen walls, ancestral shrines, stages, gables, temples and monumental archways. It is exquisitely carved and rich in themes.

• 黄山西递的门楼砖雕

Brick Carving on the Gate Tower of Xidi, Huangshan City

• 北京北海公园琼岛善因殿的琉璃砖雕佛像
Glazed Buddhist Statue in the Shanyin Hall of the Jade Islet in Beihai Park, Beijing

• 山西王家大院檐下的砖雕
Brick Carving under the Eaves of the Wang's Courtyard in Shanxi Province

• 墙上的植物图案砖雕
Plant Patterns of Brick Carving on the Wall

木雕

建筑装饰中的木雕主要应用于木结构的房顶、梁枋、护栏、室内家具等，利用不同的木材质感进行雕刻加工，丰富建筑形象。据考证，殷商时期，木雕就已在建筑中出现。现存的建筑装饰木雕集中体现在明清时期的宫殿、寺庙、民居、戏台、祠堂等之上。这一时期，木雕几乎遍及了建筑的藻井、梁枋、斗檐柱、门、窗等各个建筑构件，题材多样，工艺精湛。

浙江东阳木雕、浙江乐清黄杨木雕、广东潮州金漆木雕、福建龙眼木雕被称为“中国四大木雕”。

Wood Carving

Wood carving in architectural decoration is mainly used on wooden roofs, girders, guard rails and interior furniture. Different types of woods are used in order to enrich the image of the architecture. According to the research, during the Shang Dynasty, wood carving has been applied in buildings. The currently preserved wood carvings are mainly on palaces, temples, residential buildings, stages and ancestral shrines of the Ming Dynasty and the Qing Dynasty. During the period, wood carving was prevalent on the caisson, girder, eaves column, doors, windows and other building components themed on various subject matters and exquisitely engraved.

Wood carving of Dongyang in Zhejiang Province, boxwood carving of Yueqing in Zhejiang Province, gold foil wood carving of Chaozhou in Guangdong Province and longan wood carving of Fujian Province are known as the "Top 4 Chinese Wood Carvings".

• 龙凤呈祥木雕
Prosperity Brought by the Loong and the Phoenix

四大木雕

浙江东阳木雕约始于唐而盛于明清，自宋代起已具有较高的工艺水平，主要用于宫殿、寺庙、园林、住宅等建筑的装饰。东阳木雕以平面浮雕为主，有薄浮雕、浅浮雕、深浮雕、高浮雕、多层叠雕、透空双面雕、锯空雕、满地雕、彩木镶嵌雕、圆木浮雕等类型。东阳木雕作品一般不加彩绘，而是保留原木的天然纹理和色泽，格调高雅，又称“白木雕”。题材内容多为历史故事和民间传说。

浙江乐清黄杨木雕以黄杨木为雕刻材料，创始于宋元，流行于明清。题材内容大多为圆雕的八仙、寿星、关公、弥勒、观音等民间神话传说中的人物。

潮州金漆木雕最初是中国古代的一种建筑装饰艺术形式，之后逐渐形成一种木雕流派，多以樟木或杉木为雕刻材料，加以生漆和金箔。雕刻形式有浮雕、立体雕、通雕等。

福建龙眼木雕因雕刻材料为龙眼木而得名。龙眼木雕以圆雕为主，也有浮雕、透雕。题材内容多为寿星、渔翁、弥勒、达摩、仙女等，也有草虫、花卉、果盘、牛、马、熊、狮、虎、金鱼、仙鹤等。

• **浙江东阳木雕**
Wood Carving of Dongyang, Zhejiang Province

- 黄杨木雕刘海戏金蟾

Boxwood Carving Liu Hai and the Toad

Top 4 Chinese Wood Carvings

The wood carving of Dongyang in Zhejiang Province began in the Tang Dynasty and prevailed in the Ming and Qing dynasties (1368-1911). A high level of workmanship was achieved since the Song Dynasty. It was mainly used for manufacturing architectural decoration in palaces, temples, gardens, residential houses and other buildings. Dongyang wood carving is mainly flat relievo. It includes thin relievo, low relievo, deep relievo, high relievo, multi-layered carving, hollow double-sided carving, pierced carving, all-over carving, color wood inlay carving and log relievo. Generally speaking, work pieces of Dongyang wood carving are not painted in order to retain the natural texture and color of the wood. It is elegant in style and is also known as "white wood carving". Its themes and contents are mainly historical stories and folklore.

The boxwood carving of Yueqing in Zhejiang Province dates back to the Song and Yuan dynasties (960-1368) and prevailed in the Ming and Qing dynasties. Like circular carving, its themes and contents are mainly figures in folk myths and legends such as the Eight Immortals, the God of Longevity, Guan Gong (a famous general in the Three Kingdoms Period (220-280) renowned for his bravery and code of brotherhood), Maitreya and Avalokitesvara.

The gold foil wood carving of Chaozhou, Guangdong Province, was originally an architectural decoration of ancient China. Later on, it gradually developed into a school of wood carving. Its main carving materials are camphor wood and cedar. They were coated with Chinese lacquer and gold foil. Its forms were relief, three-dimensional carving and hollowed-out carving.

The longan wood carving of Fujian Province was named after the carving material "longan wood". Its main form is circular carving, but also has forms such as relievo and openwork carving. Its themes and contents are mostly the god of longevity, fishermen, Maitreya, Bodhidharma, fairies, or grass, insects, flowers, fruit, as well as cattle, horses, bears, lions, tigers, goldfish and cranes.

• 清末金漆木雕隔扇门
Gold Foil Wood Carving of Partition Door in the Late Qing Dynasty

陶塑

中国很多地区都有用陶塑装饰屋脊的风俗，例如云南昆明农村的民居正中上方房顶常装有瓦猫，用来保祐全家平安。

Pottery Sculpture

Many regions in China have the custom of decorating roof ridges with pottery sculptures. For example, in the rural areas of Kunming City in Yunnan Province, a tile cat is positioned on the roof of the upper middle area of a house to protect the whole family.

- **造型夸张的瓦猫**

瓦猫为蹲坐式，与一块盖瓦塑在一起。瓦猫的形象并不统一，有的像虎，有的像麒麟，还有的被塑成带飞翼的虎，并附带有各种象征吉祥的物件。

Exaggeratedly-designed Tile Cat

The tile cat is in a squatting posture and molded together with a cover tile. The tile cats don't have a uniform image. Some look like a tiger while some like a kylin. Others are molded into tigers with flying wings and supplemented with various mascots.

• 屋顶陶塑装饰
Pottery Decoration on the Roof

灰塑

灰塑是在福建、广东地区广为流行的一种屋顶装饰艺术，也叫“灰批”，就是用白灰和蚝壳灰为原料制成灰膏，然后用捏塑的方法做出丰富多彩的雕塑。灰塑可以制成浮雕或圆雕作品。灰塑的题材十分丰富，包括历史人物、吉祥题材、戏文故事等内容。在灰塑上加上五彩的颜色称为“彩描”。彩描是一种画塑结合的方法，先用灰塑技法做出装饰纹样，然后再绘以彩画，常用来装饰屋脊和山墙部分。

Limestone Sculpture

Limestone sculpture is a widely-popular art form of roof decoration in Fujian Province and Guangdong Province, also called "*Huipi*". Emplaster is made by mixing the raw materials of lime and oyster-shell lime mortar and then kneaded by hand to make various colorful sculptures. Limestone sculptures can be made into works of relievo or circular carving. It has a wide range of themes, including the representation of historical figures, auspicious themes, skits and stories. Limestone sculptures can include

陈家祠堂屋顶彩描色彩鲜艳，雕刻精美，屋脊上的瑞兽栩栩如生。

Colorful and exquisitely engraved color depiction on the roof and the lively auspicious animal on the roof ridge of the ancestral shrine of the Chen's Courtyard.

various colors and be turned into what is called "color depiction", which is a method of combining paintings with sculptures. First, decorative patterns are made adopting the technique of limestone sculpture and then they are painted with colors. Color depiction is usually used to decorate the roof and the gable wall.

• 广州陈家祠堂屋顶彩描

Color Depiction on the Roof of the Ancestral Shrine of the Chen's Courtyard in Guangzhou

屋顶陶塑装饰之灰塑
Limestone Sculpture of Roof-top Pottery Decoration

彩绘

传统建筑的彩绘在中国有悠久的历史，是传统建筑装饰中最突出的特点之一，以独特的风格、精湛的制作技术及富丽堂皇的装饰艺术效果闻名于世。“雕梁画栋”这一成语就是用来形容建筑物富丽堂皇的。

彩绘原本是为了防止木结构潮湿、腐坏、虫蛀的；后来，其装饰性才开始凸显出来。宋代以后，彩绘已成为宫殿不可缺少的装饰艺术；清代则是中国传统建筑彩绘发展的全盛时期。彩绘有三大类：和玺彩画、旋子彩画和苏式彩画。

和玺彩画是清代宫廷建筑上最

Color Painting

Color painting on traditional architecture has a long history in China and is one of the most prominent features of traditional architectural decoration. Its unique style, superb manufacturing techniques and magnificent art effect are world-renowned. The idiom "carved beams and painted rafter (*Diaoliang Huadong*)" is used to describe a magnificent building whose beauty is attributable to color painting.

Color painting was originally adopted to prevent humidity, rot and bristletail, and used for decorative purposes later on. After the Song Dynasty, color painting has become an indispensable decorative art for palaces. The Qing Dynasty marked the prime

高等级的彩画，产生于明末清初。其特点是枋心以括线括起，枋心内画以龙、凤等主纹。制作和玺彩画的主体框架线路一律采用沥粉、贴金做法，不采用勾墨线的方法；细部纹饰大部分也以沥粉、贴金方法制成，故具有金碧辉煌的效果。因为宫廷建筑有严格的等级之分，所以和玺彩画也以纹样分为不同的使用等级。

枋心是檩枋中心，是梁枋彩画的中心部位。藻头俗称“找头”，

• 华丽的和玺彩画
Gorgeous Loong Pattern

time of the development of traditional Chinese architectural color painting. Color painting can be classified into three categories: loong pattern, tangent circle pattern and Suzhou-style pattern.

Appearing in the late Ming and early Qing dynasties, loong pattern marks the highest level of color painting on the imperial buildings of the Qing Dynasty. Its feature is to attach vinculum to and draw patterns of loong and phoenix on the central portion of the painted beam. The main frame of the loong pattern adopts the method of gelled patterning and gilding instead of ink lines. Most of the detailed patterns adopt the same method, achieving a magnificent effect. Imperial buildings have a strict hierarchy, so loong patterns are divided into different application levels.

The central portion of a painted beam is the center of girder color painting. The intermediate portion of painted beam refers to the part from the end to the center of the painted beam and is comprised of thong line, box, the intermediate portion and the end portion of the painted beam. The thong line is one of the five lines and a component of the intermediate portion of the painted beam. The box is a small space of the

指檩端至枋心的中间部位，由找头本身、皮条线、盒子、箍头等部分组成。皮条线是五大线之一，亦是组成找头的一个部分。盒子是找头部分的一段小空间。箍头是檩枋尽端处的彩绘线。

intermediate portion of the painted beam. The end portion of the painted beam is the color painting line at the end of the purlin.

北京颐和园金龙和玺彩画

金龙和玺彩画是最高级别的彩画，只用于皇帝登基、理政的殿宇和重要的坛庙主殿。枋心以龙纹为主，以云气、火焰等纹饰为辅，枋心画以二龙戏珠，藻头中青地画以升龙，绿地画以降龙，盒子中画以坐龙，藻头较长时画以双龙。

Golden Loong Pattern in the Summer Palace, Beijing

The golden loong pattern is the highest level of color painting exclusively used in palaces for emperors ascending the throne and administrating the country and in the main halls of important altars and temples. The central portion of the painted beam is mainly engraved with a loong pattern supplemented by other patterns like cloud, gas and flame. The central portion of the painted beam has the drawing of "two loongs play with a pearl" pattern; a flying loong was drawn on the blue space of the intermediate portion of the painted beam and a descending loong on its green space; the pattern of a squatting loong was drawn in the box and twin loongs are painted when the intermediate portion of the painted beam is long.

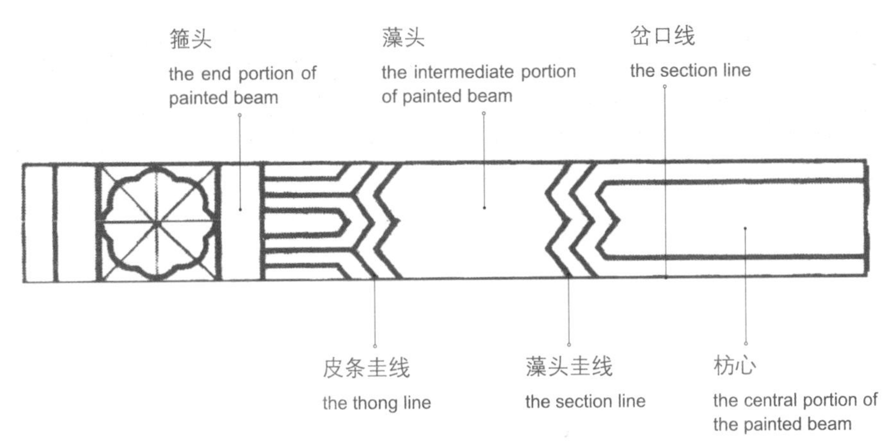

和玺彩画纹样的结构示意图
Diagram of Structure of Loong Pattern

北京天坛祈年殿龙凤和玺彩画

龙凤和玺彩画级别稍低，主纹是龙、凤两种纹样，一般在青地画龙，在绿地画凤。龙凤和玺彩画中有龙凤呈祥、双凤昭富等纹饰。龙凤和玺彩画用于帝后寝宫和祭天的主殿，如天坛祈年殿。

Loong and Phoenix Pattern in the Hall of Prayer for Good Harvests of the Temple of Heaven, Beijing

The loong and phoenix pattern is of a relatively low level with the loong and phoenix being the main pattern. Usually, the loong is painted on the blue space and the phoenix on the green space. There are characters of "prosperity brought by the loong and the phoenix" and "two phoenixes bringing about wealth". The loong and phoenix pattern is used in the sleeping palace of the emperor and the empress and the main hall for worshiping the heaven, such as the Hall of Prayer for Good Harvests in the Temple of Heaven.

• 北京故宫体仁阁的龙草和玺彩画

龙草和玺彩画级别低于龙凤和玺彩画，主纹是龙纹和大草纹，在绿地画龙，在红地画草。龙草和玺彩画用于皇宫的重要宫门，主轴线上建筑的配殿、配楼，以及重要寺庙的殿堂。

Loong Grass Pattern in the Belvedere of Embodying Benevolence of the Forbidden City, Beijing

The level of the loong grass pattern is lower than the loong pattern. Loong and grass are the main patterns with the loong pattern being drawn on the green space and the grass pattern on the red space. The loong grass pattern is used on important imperial gates of the palace, the side halls and side buildings of architecture on the main axis and the halls of important temples.

旋子彩画因藻头内画以旋涡状的旋子图案而得名，其等级低于和玺彩画。明代时，旋子彩画就已被使用；清代时，它一般用于官署和寺庙的主殿、配殿，牌楼等建筑。枋心部分和盒子部分的彩画内容可随着等级的高低而变化，从高到低依次是龙纹、龙凤纹、凤纹、锦纹、夔龙纹、卷草纹和花卉纹等。

Named after the spiral roton pattern inside the intermediate portion of the painted beam, the tangent circle pattern has a lower rank compared with the loong pattern. In the Ming Dynasty, the tangent circle pattern has been adopted. In the Qing Dynasty, it was usually seen on the main halls, side halls and decorated archways of government offices and temples. The content of pattern of the central portion of the painted beam and the box can change as the grade levels differ. The levels from high to low are loong pattern, loong and phoenix pattern, phoenix pattern, mosaic pattern, *Kui* loong pattern, curly grass pattern and flower pattern.

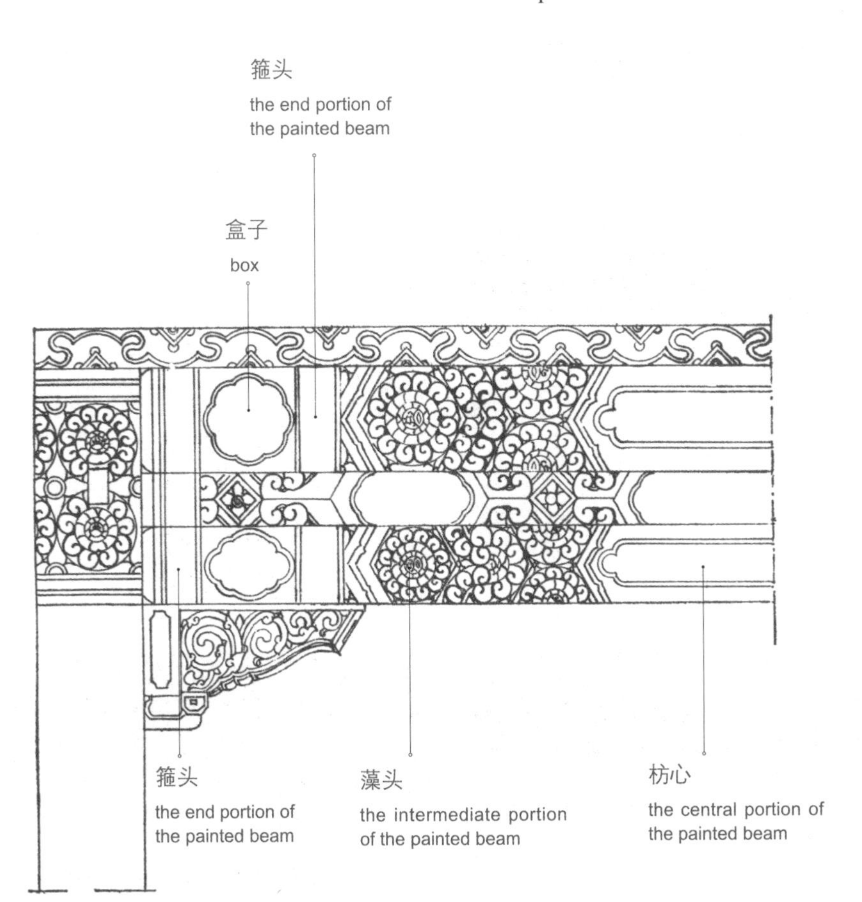

• 旋子彩画纹样的结构示意图
Diagram of Structure of Tangent Circle Pattern

• 各式各样的旋子彩画藻头纹

Various Tangent Circle Patterns on the Intermediate Portion of the Painted Beam

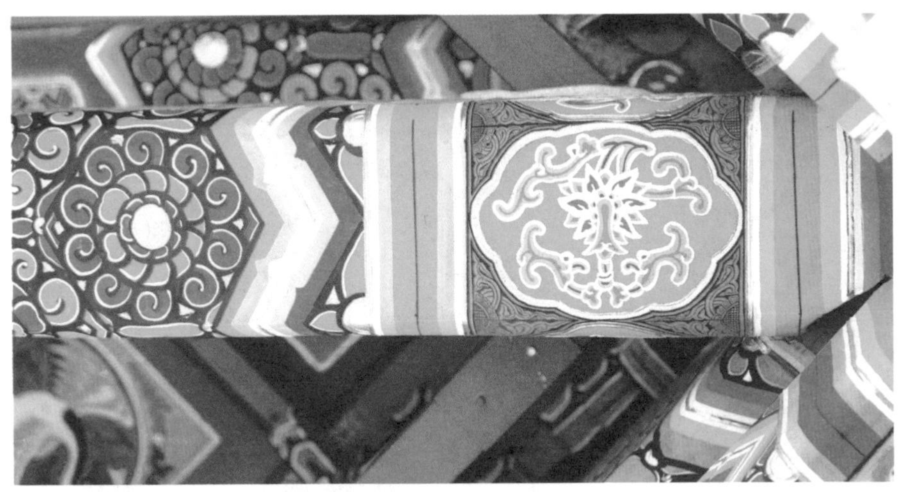

- **旋子彩画的盒子纹**

盒子是位于箍头之间的装饰纹样单元，因纹样的轮廓像一个盒子而得名。

Tangent Circle Pattern, Box Pattern

Box is a unit of decorative pattern between the end portions of painted beams. It is named due to its resemblance to a box.

- **旋子彩画的箍头**

Tangent Circle Pattern, the End Portion of the Painted Beam

• **旋子彩画的旋花**

旋花是构成旋子彩画的主要图案，表现为在藻头内以旋涡状的几何图形构成一组圆形的花纹图案。因最外一层花瓣呈旋涡状，故名“旋花”。

Tangent Circle Pattern, Spiral Flower

The spiral flower is the main component of the tangent circle pattern. A series of round flower patterns is formed using spiral geometric figures inside the intermediate portion of the painted beam. The outer layer of the petal is spiral, therefore it is called "spiral flower".

旋子彩画也有等级之分，以用金量的多少作为分等级的依据。其中浑金旋子彩画在旋子彩画里是等级最高的，彩画部分被全部贴以金箔，不施以其他颜色。然后是金琢墨石碾玉旋子彩画。

The tangent circle pattern can also be graded according to different gold content. The tangent circle pattern of gold ranks the highest because its color painting is completely covered with gold foil instead of other colors. The tangent circle pattern of gelled patterning and gilding ranks the second.

• 北京故宫三大殿西庑房墨线大点金旋子彩画
Ink Line Gold Tangent Circle Pattern in the West Wing Room of the Three Grand Halls (the Hall of Supreme Harmony, the Hall of Central Harmony and the Hall of Preserving Harmony) in the Forbidden City, Beijing

苏式彩画源于江南的苏州一带，明清时期传至北方并进入宫廷，成为官式彩画中的一个重要品种。苏式彩画主要用于园林中的小型建筑——亭、台、廊、榭等，以及四合院、垂花门的额枋上。南方气候潮湿，彩画通常只用于内檐，外檐以砖雕或木雕装饰；北方则内外兼施。

The Suzhou-style color painting originates from the Suzhou area of the south of the lower reaches of the Yangtze River. During the Ming Dynasty and the Qing Dynasty, it was moved north to the palace and became an important type of palatial color painting. Suzhou-style color painting is mainly used for small architecture in gardens—pavilions, platforms, colonnades and pavilions on terraces etc., as well as the architrave of quadrangle dwellings and festooned doors. Since the climate in the south is humid, color painting is normally used on the inner brim. Decoration on the outer brim is brick carving or wood carving. However, color painting is adopted inside and outside the brim in the north.

• **北京颐和园建筑小额枋上的苏式彩画双燕图**

宽大的荷叶上方以红、白、绿色晕染出开得正艳的荷花，在荷的衬托下，双燕形象显得更加生动。

Suzhou-style Color Painting of Twin Swallows on the Architrave in the Summer Palace, Beijing

Above the large lotus leaf lies lotus flowers in full blossom painted in red, white and green. The lotus flowers set off the vivid image of twin swallows.

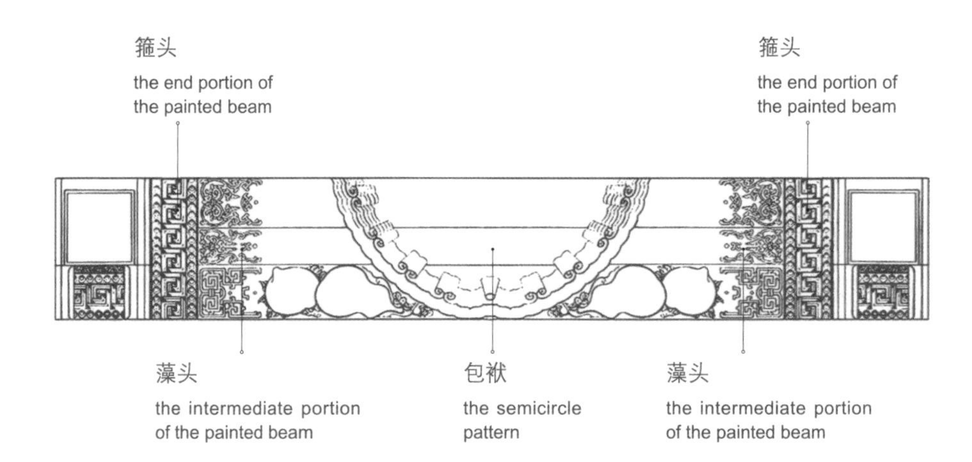

• **苏式彩画纹样的结构示意图**

Diagram of Structure of Suzhou-style Color Painting

• 苏式包袱彩画

北方内檐苏画与和玺、旋子彩画相同，采用狭长枋心。在外檐部分，檩、垫、枋三部分枋心常连成一体，形成一个大的半圆形的“搭袱子”，俗称“包袱”。根据内部彩画题材的不同，包袱可分为“花鸟包袱”“人物包袱”“线法套景包袱”等。

Suzhou-style Semicircle Color Painting

Like the loong pattern and the tangent circle pattern, Suzhou-style color painting in the north adopted a narrow central portion of the painted beam on the inner brim and combines purlin, bolster and beam on the outer brim to form a large semicircle pattern. It can be classified into various types such as the flower and bird pattern, the figure pattern and the scenery pattern according to different painting themes.

• 北京故宫养心殿套景包袱苏式彩画

慈禧太后喜欢苏式彩画，内廷东西六宫、宁寿宫区的一些宫殿上均绘以苏式彩画。

Suzhou-style Semicircular Scenery Pattern in the Palace of Mental Cultivation in the Forbidden City, Beijing

Empress Dowager Cixi liked Suzhou-style color painting. Therefore, the entire six western and eastern palaces of the inner court and some imperial residences surrounding the Palace of Tranquil Longevity adopt Suzhou-style color painting.

• 苏式彩画之山水套景包袱
Semicircle Pattern of Scenery in Suzhou-style Color Painting

• 苏式彩画之人物套景包袱
Semicircle Pattern of Figures in Suzhou-style Color Painting

苏式彩画之花鸟包袱

Semicircle Pattern of Flowers and Birds in Suzhou-style Color Painting

苏式彩画之人物包袱

Semicircle Pattern of Figures in Suzhou-style Color Painting

• 北京恭王府花园苏式彩画

五福捧寿图为苏式彩画中最华贵的一种形式——金琢墨，工艺考究，用金多，画面精致。

Suzhou-style Color Painting in the Garden of Prince Gong's Mansion, Beijing

The diagram of "five bats holding longevity" is the most luxurious form of Suzhou-style color painting. It uses a large amount of gold to produce a delicate image with sophisticated techniques.

• 海墁苏画

海墁苏画多用于次要部位，无枋心、包袱，表现为在梁枋的箍头或卡子之间画一些简单的花纹，是苏式彩画中的一般画法。

Haiman Suzhou-style Color Painting

Haiman Suzhou-style color painting is mainly used on less important parts such as the end portion of a painted beam without a central portion, a semicircle color painting or a girder and space between clamps. Usually, it is painted in simple patterns and is a general type of painting technique in Suzhou-style color painting.

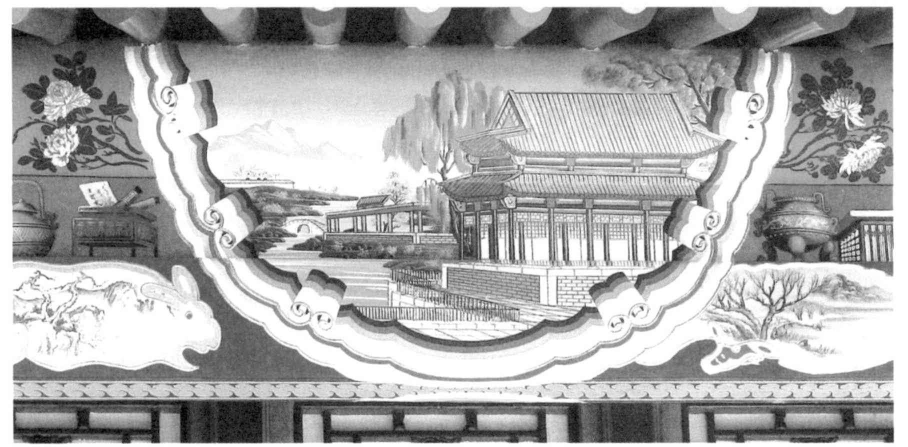

• **金线苏画**

金线苏画是最常见的苏式彩画，箍头线、包袱线等均被沥粉、贴金。

Golden Thread Suzhou-style Painting

The golden thread Suzhou-style painting is the most common Suzhou-style color painting. Gelled patterning and gilding are used on the lines for the end portion of the painted beam and for the semicircular pattern of the painting.

彩画不仅仅是一种文化艺术，更是一门需要极高技巧的工艺。建筑彩画装饰的基本步骤包括基层处理、做地仗、刷油漆、作画四部分。不论是建筑中的桁、梁、枋、门板、隔扇、窗棂，还是被精雕的斗拱、雀替，所使用彩绘的技巧大致分为以下四种。

沥粉、贴金：沥粉就是以压挤粉浆的方式突出图案的线条。贴金就是用胶油把金箔贴在檐梁或木刻品上，有沥粉即有贴金，用金量越

Color painting is not only a cultural art, but also a craftsmanship requiring high skills. The four basic steps of adopting color painting as architectural decoration are preparing the foundation, making the ground layer, painting and drawing. Be it a joist, girder, beam, door plank, partition board, window lattice or exquisitely-carved arch bracket and sparrow brace, the techniques of color painting used can be generally classified into the following four types:

Gelled patterning and gilding: gelled patterning refers to a technique of

多，越能显示建筑的价值。

平涂法、色晕法：门板、隔扇等不画图案的部分大多以大笔平涂的方式进行着色，要求色彩均匀。色晕指用不同的深浅来表现色彩，形成渐变的效果。

五彩遍装：即依照不同图案的需要，使用胭脂、槐花、石青、石绿、朱砂等植物颜色，给木结构建筑披上华丽的外衣。

水墨技法：即在传统建筑的墙壁上题字、作画。

squeezing slurry to highlight the lines of the image; gilding refers to the technique of sticking gold foil to beam or wood carving. Gelled patterning cannot exist without gilding. The gold content reflects the value of the architecture.

Techniques of flat painting and color-gradation: the technique of flat painting is normally used on spaces that do not need pictures, such as on a door and partition board. Color-gradation refers to the technique of expressing shades of colors to achieve an effect of gradational color stratification.

All-around color painting: based on the needs of different images, plant colors of rouge, pagodatree flower, azurite, malachite and cinnabar are used to cover the wooden architecture with a gorgeous coat.

Ink painting technique: refers to the technique of inscribing and drawing on the walls in traditional architecture.

传统建筑外部装饰

The Exterior Decoration of Traditional Architecture

中国人向来注重“天时、地利、人和”，崇尚自然，并与之和谐相处，中国传统建筑直接与外部环境相关联的外部结构很好地体现了这一点。传统建筑的外部结构包括屋顶、墙、门窗、梁柱、台基、地面等部分，这些组成部分的装饰具有直观、鲜明的特征。

Chinese people always attach great importance to "favorable climatic, geographical and human conditions". They respect nature and live in harmony with it. As a direct link with the exterior environment, the exterior structure of traditional Chinese architecture greatly reflects the harmony between man and nature. The external structures of traditional Chinese architecture including roofs, walls, doors, windows, beams, pedestals and grounds, have a direct and distinctive feature in terms of the decoration.

> 屋顶装饰

屋顶上的建筑装饰是传统建筑装饰中最重要的部分，具有很高的艺术价值。针对不同类型、不同等级的建筑，屋顶的装饰也有所不同。脊兽、瓦当、滴水、博风板、悬鱼、陶塑、灰塑、惹草等都是传统建筑屋顶的主要装饰构件。

脊兽

脊兽是被设置在中国传统建筑屋顶的屋脊上的兽形构件，不同位置的脊兽有着不同的功能和寓意。传统建筑屋顶上安装的脊兽的用途主要有三种：第一种是古人为求吉利而设的吉祥物；第二种是对官式建筑而言，屋顶吻兽数量的多少代表了殿宇等级的高低；第三种是防

> The Roof Decoration

The roof-top architectural decoration is the most important part and owns great artistic value. For different types and levels of buildings, the roof-top decoration is different. The ridge animal, eaves tile, drip channel, bargeboard, suspended fish, pottery decoration, plaster sculpture and triangle board are all major components of roof-top decoration in traditional architecture.

The Ridge Animal

The ridge animal is a decorative object installed on the ridge of traditional Chinese architecture. It has different functions and meanings when deployed in different places. Ridge animals on the roof of traditional Chinese architecture mainly have three functions: the first is to bring about good luck; the

水，脊兽本身就是被专门设计、制作的防水构件，如有的脊兽被设置在正脊与四条垂脊的交会点上，而此处正是屋顶防水最不好处理的位置，脊兽起到了封固的作用，能够延长建筑物的寿命。

蹲兽也称“走兽”“仙人走兽”，是被设置在宫殿建筑庑殿顶的垂脊上或歇山顶的戗脊前端的脊兽，分仙人和走兽两部分，其数量与宫殿的等级相关。每一个兽都有自己的名字和作用。还有一种走兽名为“跑龙”，主要被用在神庙正脊和亭子的垂脊上。

second is to reflect the rank of official buildings judging from the number of zoomorphic ornaments on the roof; the third is to guard against water. The ridge animal itself is specially designed and manufactured to be water-proof. For example, some ridge animals are installed on the intersection of the main ridge and the four vertical ridges where it is the most difficult to resist water. The ridge animal serves the function of fixing and sealing, hence extending the service life of the building.

Squatting animals, also known as "running animals" and "immortal animals", are ridge animals installed in front of the vertical ridge of hipped roof palatial buildings and the angle ridge of gable roof palatial buildings. They can be classified into immortal beings and running animals and their number is related to the rank of a palace. Each animal has its name and function. There is also a kind called running loong which is mainly used on the main ridge of a temple and the vertical ridge of a pavilion.

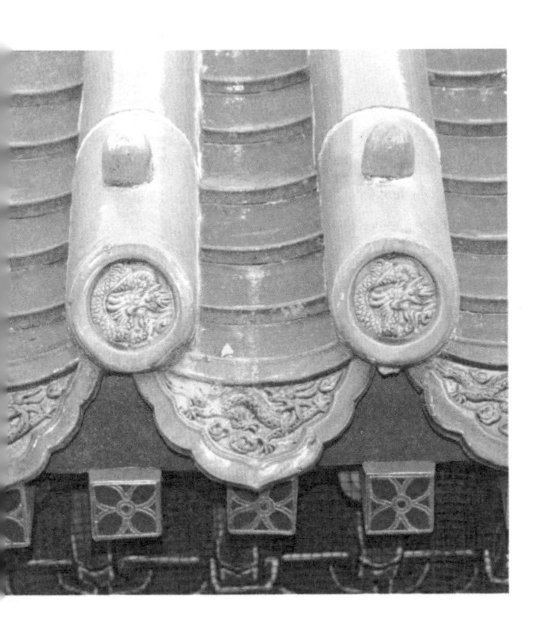

• 屋顶装饰
Roof Decoration

承德须弥福寿寺殿顶铜跑龙

河北承德须弥福寿寺的铜质屋顶上有四条斜脊，共有八条铜铸跑龙。

Copper Running Loong's on the Roof of the Sumeru Longevity Temple, Chengde

There are four sloping ridges on the copper roof of the Sumeru Longevity Temple in Chengde City, Hebei province, and eight copper running loongs lying on it as decoration.

北京故宫太和殿

The Hall of Supreme Harmony in the Forbidden City, Beijing

- **北京故宫太和殿蹲脊兽**

清代规定，骑凤仙人之后的蹲脊兽的数量为奇数，九为最高，如中和殿、保和殿、乾清宫，都排列了九个走兽。但为了凸显太和殿是故宫内等级最高的建筑，又增加了一个行什，这在中国宫殿建筑史上是独一无二的。

Squatting Ridge Animal of the Hall of Supreme Harmony in the Forbidden City, Beijing

The Qing Dynasty (1616-1911) issued an order that the number of squatting ridge animals behind the phoenix-riding immortal should be odd and nine at most. For example, the Hall of Central Harmony, the Hall of Preserving Harmony and the Palace of Heavenly Purity all installed nine running animals. However, in order to highlight that the Hall of Supreme Harmony is the highest ranking building in the Forbidden City, a decorative little animal called "*Hangshi*" is added, something unprecedented in the history of Chinese palatial buildings.

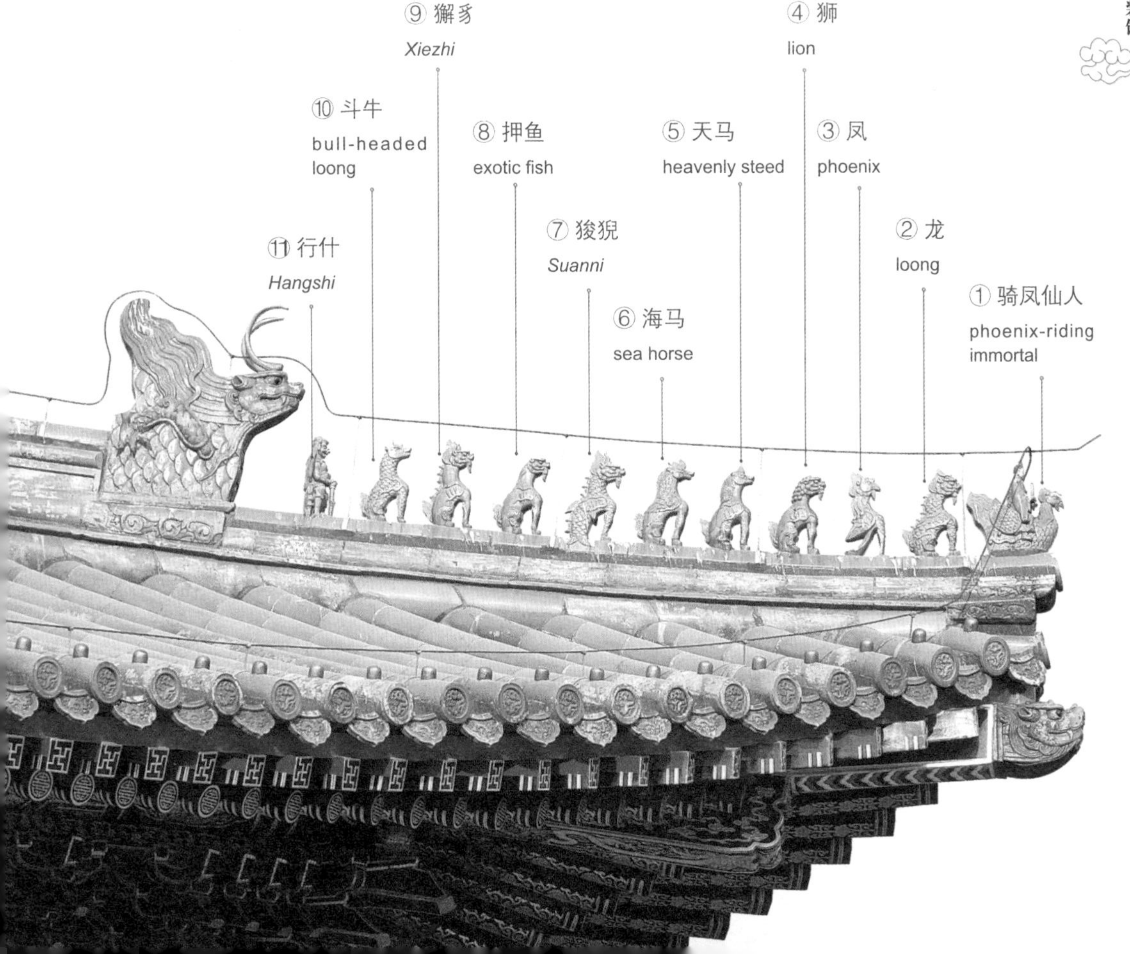

① 骑凤仙人：被放于顶端，意为无路可走，只能骑凤飞行。

Phoenix-riding Immortal: placed on the top, meaning there is no alternative way other than flying a phoenix.

② 龙：传说中的神奇动物，能兴云作雨，是吉祥兽。

Loong: a magical animal in Chinese legend that can summon clouds and rain. It is an auspicious animal.

③ 凤：即凤凰，传说中的鸟王，是吉瑞的象征。

Phoenix: king of birds in Chinese legend and a symbol of good luck.

④ 狮：传说中狮是兽中之王，是威武的象征。

Lion: king of animals in Chinese legend and a symbol of might.

⑤ 天马：神马，有翅膀，可在天上飞行，象征皇家的威德可通天庭。

Heavenly Steed: a divine horse that has wings and can fly. It symbolizes that the might and virtue of the royal family can reach Heaven.

⑥ 海马：名“落龙子”，造型近似天马，但无翅膀，被饰以火焰。海马象征皇家威德可达海底。

Sea Horse: known as the “fallen loong”, is an animal resembling the heavenly steed. Instead of having wings, it has blaze as its decoration. It symbolizes that the might and virtue of the royal family can reach the sea bottom.

⑦ 狻猊：传说中能食虎和豹的一种猛兽，形象类狮，有威武、百兽率从之意。

Suanni: a wild beast that can eat tigers and leopards in Chinese legend and resembles a lion. With its incredible might, it is the leader of all animals.

⑧ 押鱼：传说中的海中异兽，身披鱼鳞，有鱼尾，可兴云作雨、灭火防灾。

Exotic Fish: an exotic sea animal in Chinese legend with fish scales and a tail. It can summon clouds and rain. Its magic also includes fighting fire and preventing disasters.

⑨ 獬豸：传说中的独角猛兽，善辨是非和曲直，古时以它作为皇帝"正大光明""清平、公正"的象征。

Xiezhi: a legendary kylin beast that is able to distinguish between right and wrong. In ancient times, it was a symbol of the fairness, honesty and justice of the emperor.

⑩ 斗牛：传说中的一种龙，牛头兽造型，身披龙鳞，遇阴雨作云雾，常位于道旁及金鳌玉栋坊之上，也是人们所认为的能除祸、灭灾的吉祥物。

Bull-headed Loong: a legendary loong with a bull head and loong scales. It gathers clouds to generate rain and is usually found on the roadside or the Golden Turtle and Jade Rainbow Lane. It is also considered to be a mascot to prevent disasters.

⑪ 行什：传说中的一种带翅膀、有着猴面孔的"神仙"，挺胸凸肚，手执金刚杵，似在执行监押的任务。《清式营造则例》称之为"猴"。

Hangshi: a legendary God with wings and a monkey face. It thrusts out its chest and holds a Vajra Pestle as if performing the duty of guarding the inmates. It is referred to as "monkey" in the *Building Standards of the Qing Dynasty*.

• **北京故宫雨花阁殿顶跑龙**

中国古代重要的殿宇建筑物上还有跑龙装饰。故宫雨花阁的屋顶是铜顶，屋顶的4个转角上有4条跑龙。据清宫档案记载，4条跑龙共用铜720斤。民间礼制庙宇也有用跑龙进行装饰的。例如广东五华县文庙的大成殿，殿顶正脊上有两条跑龙相对，曲身相斗，生动异常。

Running Loong of the Tower of Raining Flowers in the Forbidden City, Beijing

There are running loong decorations on important ancient Chinese palatial buildings. The Tower of Raining Flowers of the Forbidden City has a copper roof and four running loongs on the four corners of the roof. According to the archives of the Qing Dynasty, the four running loongs used a total of 720 *Jin* (360 kilograms) of copper. There are also private ritual temples using running loongs as decoration. For example, the Dacheng Hall of the Confucius Temple in Wuhua County, Guangdong Province has two running loongs oppositely placed on the main ridge of its roof. It is so lively that it looks as if the two loongs are fighting each other.

垂兽又称“角兽”，是垂脊上的兽头形构件，位于蹲兽的后面。垂兽多以琉璃瓦制成，长有双角，中间被掏空，用来钉入垂兽桩加以固定。其作用是防止垂脊上的瓦下滑或被吹落，避免雨水渗漏，并加固与屋脊相交的部位。

Suspended beast, also known as "horned beast", is a decorative beast-head-shaped object on the vertical ridge sitting behind the squatting beast. Most suspended beasts are made of glazed tiles and have two horns. The middle of it is hollowed out for the fixing stake. The function of the suspended beast is to

• 琉璃垂兽
Glazed Suspended Beast

戗兽是戗脊上的兽形构件，用于歇山顶和重檐建筑上。戗兽与垂兽的造型大体相同，都是螭龙之形，但比垂兽略小一些，俗称“兽头”。戗兽将戗脊分为“兽前”和

prevent tiles on the vertical ridge from falling, avoid rainwater leakage and consolidate the intersection of roof ridges.

Hip-mounted beast is an ornament on the hip ridge and used on gable and hip roof and double eave buildings. The shape of the hip-mounted beast is almost the same as that of the suspended beast which are all in a hornless loong shape. However, the hip-mounted beast is generally known as the "beast head" and is slightly smaller than the suspended beast. The hip-mounted beast divides

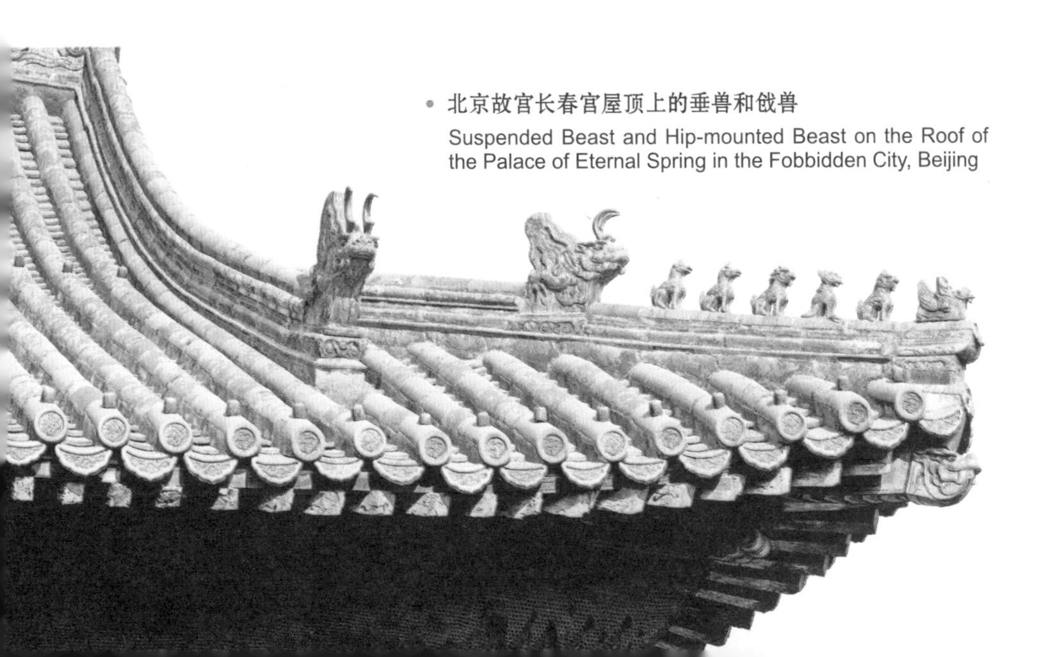

• 北京故宫长春宫屋顶上的垂兽和戗兽
Suspended Beast and Hip-mounted Beast on the Roof of the Palace of Eternal Spring in the Fobbidden City, Beijing

"兽后"。戗兽的数量设置反映严格的等级制度，数量越多，级别越高，而且数目都是奇数。奇数表示"阳"，九为最大，故建筑以九件戗兽为等级最高。

the hip roof into the front roof and the back roof. The amount of hip-mounted beasts is rigorously standardized. The more the amount, the higher the rank. All of the amounts are odd. Odd numbers symbolize "*Yang*", with nine being the largest. Therefore, a set of nine hip-mounted beasts ranks the highest.

• 西夏时期无角琉璃套兽

套兽是脊兽之一，被安装于仔角梁的端头上，作用是防止屋檐角遭到雨水侵蚀。套兽一般由琉璃瓦制成，为狮子头或龙头形状。

Glazed Hornless Beast of the Western Xia Dynasty (1038-1227)

Glazed hornless beast is one ridge animal and is installed on the end of the son corner ridge. Its function is to prevent the eaves corner from being eroded by rain. It is generally made of glazed tiles into lion-headed or loong-headed shapes.

吻兽是位于房屋屋脊两端的装饰性瑞兽，在正脊两端的被称为“正吻”，又称“大吻”“吞脊兽”等。根据形象的不同，正吻也有不同的称呼。唐至明清之前的建筑普遍采用鸱吻。鸱吻又称“螭吻”“鸱尾”，为龙头鱼身，是龙生的九子之一，可喷水降雨，人们认为它可以避免火灾。除了起装饰作用，鸱吻还可以严密封固两坡与瓦垄交会处，防止雨水渗漏。明、清两代的正吻则通常以龙吻取代鸱

Zoomorphic ornament is an auspicious beast used as decoration on the ends of the roof ridge. On the ends of the main ridge lies the "main ornament", also known as the "big ornament" or the "swallow ridge beast". The main ornament can be called different things according to different images. Architecture since the Tang Dynasty (618-907) and before the Ming Dynasty (1368-1644) and the Qing Dynasty generally adopt an owl ornament decoration which is also called "hornless loong ornament" or "owl tail-shaped ornament". It has a loong head and fish body. It is one of the nine sons of loong that can summon raims, and is supposed to avoid fire. Besides serving as a decoration, the owl ornament can also fix and consolidate the intersection of two sloping tiles to guard against rainwater leakage. In the Ming and Qing dynasties (1368-1911), an owl ornament is generally replaced by a loong-head ornament and the usage of it is strictly

龙头鱼身的鳌鱼装饰

Turtle Fish Ornament with a Loong Head and Fish Body

吻，并且对吻兽的使用有严格的规定。民居一般不被允许使用大型龙吻，只有寺庙或官宅才被允许使用，并以琉璃瓦材质为主。鳌鱼也是一种用于屋脊上的吻兽，为龙头鱼身的形象，常被置于屋顶的脊梁两端，头在下方，尾巴上扬。鳌鱼生于海里，有灭火之意，同时又喻示“独占鳌头”，即万事争第一。

regulated. Residential buildings usually were not allowed to use large loong-head ornaments except for temples and official residences where the ornament is made of glazed tiles. The turtle fish is also one of the zoomorphic ornaments used on roof ridges. The turtle fish has the appearance of a loong head and a fish body. It is usually installed on the ends of the roof ridge with its tail curling upwards on the top. It is born in the ocean and is a symbol of firefighting. It also means championship, i.e. competing to be number one.

• 北京故宫太和殿龙吻

Loong-head Ornament in the Hall of Supreme Harmony in the Forbidden City, Beijing

• 苏州拙政园远香堂正脊上的鸱吻

Owl Ornament on the Main Ridge of the Drifting Fragrance Hall in the Humble Administrator's Garden, Suzhou

龙生九子

龙是中国古代神话传说中的一种神奇的动物，是原始社会时形成的一种图腾。龙生九子即龙生了九个儿子，子子不同（九是古代的极数）。

一种说法为：

老大囚牛，为龙头蛇身，爱好音乐。

老二睚眦，为豺首龙身，好斗喜杀，是龙子中的战神。

老三嘲风，形似兽，喜欢冒险、张望。

老四蒲牢，形似盘曲的龙，喜好鸣叫、嘶吼。

老五狻猊，形似狮子，喜静而不喜动，喜欢烟火。

老六霸下，形似龟，好负重，力大无穷。

老七狴犴，形似虎，急公好义，仗义执言，能明辨是非，秉公而断。

老八负屃，身似龙，头似狮，好文，专爱书法。

老九鸱吻，龙首鱼身，口阔嗓粗，好吞，可灭火。

还有一说为：

老大赑屃，样子似龟，喜欢负重。

老二鸱吻，鱼形的龙，喜四处眺望，能够灭火。

老三狴犴，样子像虎，有威力，善狱讼。

老四蒲牢，形状像龙，但比龙小，喜欢音乐和鸣叫。

老五饕餮，形似狼，性贪吃，有首无身。

老六睚眦，为龙身有角的豺狼，喜欢争斗。

老七狻猊，形状像狮，喜欢烟火，爱好静坐。

老八椒图，形似螺蚌，好闭口，性情温顺，反感别人进其巢穴。

老九貔貅，形似狮，有嘴无肛，喜欢钱财，只进不出。

Nine Sons of the Loong

The loong is a miraculous animal in ancient Chinese myths and legends. It is a totem created during the primitive society. The Nine Sons of the Loong Legend refers to a story in which the loong had nine sons, each of them different from the others (the number "nine" in Chinese suggests a large number).

One saying is as follows:

The first son *Qiuniu* has a loong head and a snake body. It likes music.

The second son *Yazi* has a jackal head and loong body. It is bad-tempered, fractious and inclined to fight. It is the god of war for the loong sons.

The third son *Chaofeng* looks like a beast and loves to take risks.

The fourth son *Pulao* looks like a wriggling loong. It likes to roar.

The fifth son *Suanni* looks like a lion. It likes tranquility, fire and smoke.

The sixth son *Baxia* resembles a turtle. It loves carrying things, hence is mighty and strong.

The seventh son *Bi'an* is like a tiger. It stands for justice and can distinguish right and wrong and hear legal cases.

The eighth son *Fuxi* looks like a loong. Its head is like a lion's. It likes writing and calligraphy in particular.

The ninth son *Chiwen* has a loong head and a fish body. The bad-tempered beast likes swallowing things and can extinguish fires.

Another saying has it as follows:

The first son *Bixi* resembles a turtle. It loves carrying things.

The second son *Chiwen* is a fish-like loong. It likes to look out and can extinguish fires.

The third son *Bi'an* is like a tiger. It is mighty and likes hearing legal cases.

The fourth son *Pulao* looks like a loong but is smaller. It likes music and loves to roar.

The fifth son *Taotie* resembles a wolf. It is gluttonous. It has a head but no body.

The sixth son *Yazi* is a jackal with horns. It looks like a loong and likes fighting.

The seventh son *Suanni* is like a lion. It likes flames and sits quietly.

The eighth son *Jiaotu* resembles spiral shells and mussels and tends to close its mouth. It is obedient and hates others intruding into its residence.

The ninth son *Pixiu* looks like a lion. It has a mouth but no anus. It likes money and prefers receiving to giving.

望兽是一种位于房屋正脊顶端的兽形构件。与吻兽朝内吞脊不同，望兽的兽头似向外望去，故称“望兽”。望兽的等级不如吻兽高，常用于城墙上的城楼或商铺。

The overlooking beast is a roof ridge decoration installed on the top end of the main ridge of the house. Unlike the zoomorphic ornament that swallows the ridge inwardly, the overlooking beast looks out, hence its name. The status of the overlooking beast is lower than the zoomorphic ornament. It is usually used on gate towers and shops.

山花

歇山屋顶两端、博风板下的三角形部分即为山花。在明代以前多为透空式，仅以悬鱼、惹草等装饰。明代的山花则多为砖和琉璃所制。明代以后的山花多以砖、琉璃、木板等将透空部分封闭起来，其上施以雕刻作为装饰，成为进行房屋装饰的重要部位。

The Pediment

The pediment refers to the triangular structure between the two ends of the gable roof and under the bargeboard. Before the Ming Dynasty, the pediment was usually hollowed out and decorated with suspended fish and grass. During

• 封闭式山花
Enclosed Pediment

宝顶

宝顶位于攒尖屋的顶端，也称“绝脊”，多为金属或琉璃所制造。其下部为砖砌线脚，上部为圆形中空的宝珠。形状有圆形、宝塔形、束腰圆形等。宝顶有防水和装饰作用，其形状、纹饰多种多样。南方园林的亭子的攒尖顶端不放有宝珠，而放有一些葫芦、宝瓶、仙鹤等雕饰构件，形式更为丰富多样。

the Ming Dynasty, most of the pediments were made of bricks and glazed tiles. After the Ming Dynasty, it was generally made of bricks, glazed tiles and boards. The originally hollowed out pediment was sealed and decorated with carvings to serve as an important component of architectural decoration.

The Precious Dome

The precious dome is located on the tip of the pyramidal roof and is also known

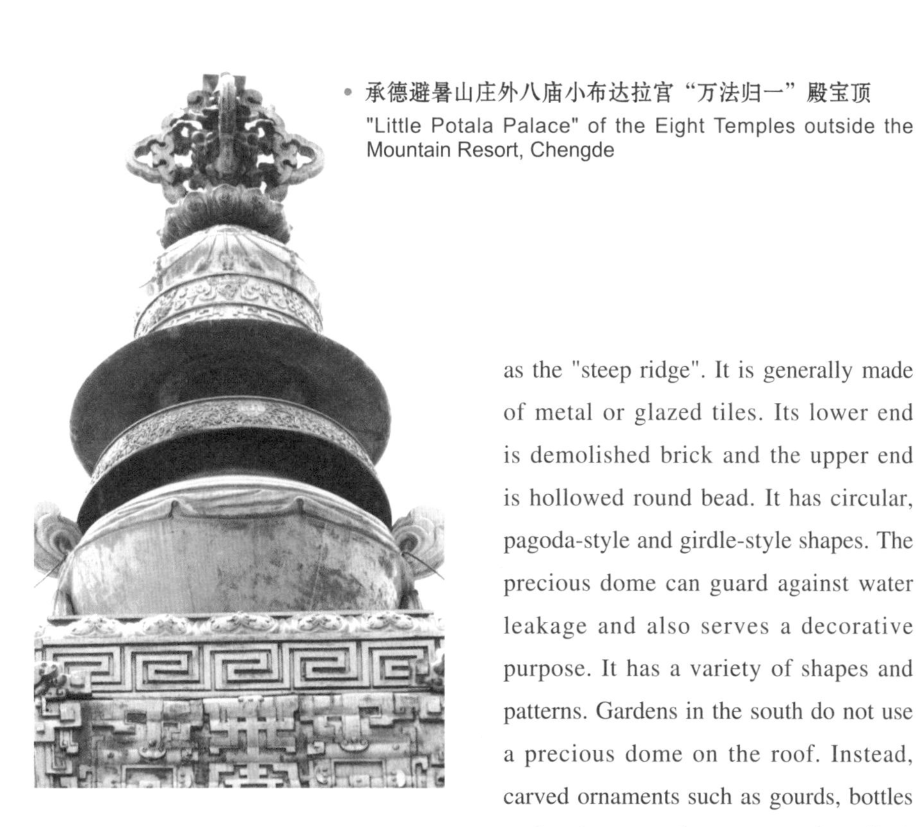

• 承德避暑山庄外八庙小布达拉宫“万法归一”殿宝顶
"Little Potala Palace" of the Eight Temples outside the Mountain Resort, Chengde

as the "steep ridge". It is generally made of metal or glazed tiles. Its lower end is demolished brick and the upper end is hollowed round bead. It has circular, pagoda-style and girdle-style shapes. The precious dome can guard against water leakage and also serves a decorative purpose. It has a variety of shapes and patterns. Gardens in the south do not use a precious dome on the roof. Instead, carved ornaments such as gourds, bottles and red-crowned cranes are installed, showing more varieties.

悬鱼

悬鱼位于建筑屋顶两端的博风板下，垂于正脊，是一种建筑装饰构件，大多以木板雕刻而成。因为最初为鱼形，从山面顶端悬垂而下，所以被称为“悬鱼”。悬鱼其实不限于鱼形，有各种各样的装饰形状，例如图像化的蝙蝠的形象，取“蝠”与“福”的谐音。悬鱼多出现在悬山顶、硬山顶的建筑中，

The Suspended Fish

The suspended fish is located under the bargeboard of the ends of the building's roof and is perpendicular to the main ridge. It is a decorative component which is mainly made of carved wood. Originally it was fish-shaped and suspended from the top of the mountain face, hence its name. In fact, it is not limited to fish shapes. It has a variety of decorative

而悬山顶、硬山顶是中国民居的常用类型，所以它多被应用在民用建筑中。

patterns such as images of bats, a name homophonic with "happiness" in Chinese pronunciation. The suspended fish usually appears on commonly used types of civil buildings such as the flush gable roof and overhanging gable roof buildings. Therefore, it can be said that most suspended fish are used on civil buildings.

博风板

博风板又叫“博缝”，位于硬山式屋顶的山墙上部，为保护屋顶侧面的木结构而设。博风板多以木板和方砖贴砌而成，有尖山式、圆山式、天圆地方式、铙钹式、琵琶式等。

• 平遥民居博风板
Bargeboard of Civil Buildings, Pingyao

Bargeboard

Bargeboard is also called "*Bofeng* (seam)". It is located on the upper part of the mountain wall of the flush gable roof building to protect the wooden structure on the side face of the roof. Bargeboard is mainly made of boards but sometimes built by laying square bricks. It can be classified into different styles such as peaked style, circular style, the "round heaven and square earth" style, cymbal style and lute style.

- **太原晋祠木悬鱼**

  悬鱼喻示水，人们认为它能避免火灾。“鱼”又与“余”谐音，喻示生活富足。

  Wooden Suspended Fish of the Jinci Temple, Taiyuan

  Suspended fish connotes water, hence it is supposed to prevent fire disasters. "Fish" is homophonic with "surplus" in Chinese pronunciation, meaning an affluent life.

瓦当

瓦当俗称“盖头瓦”“瓦头”。中国古代建筑中的陶瓦是一块压一块，从屋脊一直排列到屋檐底端的，处在众瓦之底的瓦就是瓦当，是屋顶的重要构件。瓦当的作用是保护飞檐，避免它受风吹、日晒、雨淋等，延长飞檐的使用寿命。瓦当的种类很多。根据质地可分为灰陶瓦当、琉璃瓦当和金属瓦当，形制上有半圆形、圆形和大半圆形三种，根据纹饰可分为图案纹瓦当、图像纹瓦当和文字瓦当三大类。

The Eaves Tile

The eaves tile is also called "cover tile" or "dripping tile". In ancient Chinese buildings, ceramic tiles are laid one after another from the ridge to the lower end of the eave. At the bottom of the ceramic tiles lies the eaves tile, an important component on the roof. Its function is to protect the cornice from being damaged by wind, sun and rain to extend its life. Eaves tiles can be classified into many types. For example, in terms of the texture, they can be classified into ceramic tiles, glazed tiles and metal tiles; in terms of their shape, they can be classified into semicircular, circular and large semicircular; in terms of the pattern, they can be classified into tiles of patterns, tiles of images and tiles of characters.

• **夔龙纹半瓦当**

半瓦当是早期的瓦当形式，现存最早的半瓦当为西周时期所遗存。瓦当上的夔龙造型简单，线条古朴，以直线构成夔龙纹的形象。

Semicircular Eaves Tile of *Kui* Loong Pattern

Semicircular eaves tile is a form of early eaves tile. The earliest semicircular eaves tile in existence is from the Western Zhou Dynasty (1046 B.C.-771 B.C.). It has simple *Kui* loong pattern and primitive lines. The *Kui* loong pattern is drawn with straight lines.

黄琉璃龙纹瓦当

龙首居中，龙身在圆形瓦当内部舒展，显得美观而大方。

Yellow Glaze *Kui* Loong Eaves Tile

The loong head is located in the center. The body of the loong stretches with the round tile and looks beautiful and graceful.

汉代“长乐未央”文字瓦当

Tiles of Characters "长乐未央 (Endless Joy)" of the Han Dynasty (206 B.C.-220 A.D.)

青龙纹瓦当

Eaves Tile of Green Loong Pattern

朱雀纹瓦当

Eaves Tile of Rosefinch Pattern

白虎纹瓦当

Eaves Tile of White Tiger Pattern

玄武纹瓦当

Eaves Tile of Black Tortoise Pattern

> 墙

在以木结构为主的古建筑中，墙体一般没有承重功能，只起围合、保暖、避风雨的作用。墙的装饰方式主要包括墙体堆砌和砖雕。装饰性较强的墙主要有山墙、云墙、阶梯墙、虎皮石墙、粉墙等。

> The Wall

Among ancient buildings built primarily of wood, walls generally don't serve the purpose of bearing loads but provide enclosure, warmth and shelter from the wind and rain. The decoration of walls is mainly manifested in the piling of the wall body and brick carvings. Highly decorative walls include the mountain-shape gable, cloud wall, ladder wall, tiger-skin stone walls, whitewashed wall, etc.

• 宫殿的铁红色粉墙
Iron Red Whitewashed Wall in the Palace

- **顶部覆有小青瓦且有龙头装饰的云墙**

云墙是园林中常用墙体形式之一， 又名“波形墙”，即墙头被做成波浪起伏的形状，线条流畅而轻快，富于韵律。墙面常抹以白灰，墙头覆小青瓦。

Top of the Cloud Wall Covered with Small Green Tiles and Loong Heads as Decorations

The cloud wall is commonly seen in gardens and also known as the "wavy wall" because its top is shaped like undulating waves in smooth and delightful lines with rich rhythms. The surface of the wall is often wiped with lime, while the top of the wall is covered with small green tiles.

- **虎皮石墙**

虎皮石墙是中国传统建筑常用的一种墙体，以形状不规则的石块砌成，石块间以白灰膏勾缝，有凹缝、平缝、凸缝三种勾缝的方法，既形成了自然的花纹，又可以显出石材的天然肌理效果，使墙体具有田园风格。

Tiger-skin Stone Walls

The tiger-skin stone wall is one of the most commonly used walls in traditional Chinese architecture. It is built of irregular mountain stones, with white plaster filling up the joints among the stones. The three styles of filling the joints in concave, flat and convex methods do not merely trace out natural patterns but also reveal the natural textural effects of the stone, giving the wall a pastoral touch.

• 墙头呈阶梯状错落的阶梯墙

阶梯墙又名“马头墙”，常被建于坡地，墙头呈阶梯状错落，常被做成带有凸出的线脚和小青瓦檐脊的样式。其轮廓富于变化，高低相错，有很好的装饰性，可丰富园林风景的构图。

Ladder Wall with Ladder-like Patchwork on Top

The ladder wall, also known as the "horse-head wall", is often built on slopes. The top of the wall shows a patchwork-like ladder with embossed molding lines and eave ridges of small green tiles. Its varied outlines going up and down provide itself with good decorative effects, which enrich the scenic structure of the garden landscape.

• 南方园林的白墙

White Wall of a Southern Landscaped Garden

• **北方民居的青灰墙**

粉墙又叫“混水墙”，指进行抹灰、粉刷处理后的墙体。宫廷建筑的墙多为铁红色粉墙，南方民居多为白墙，北方民居多为青灰墙。

Greenish-gray Wall of Local Residence in North China

The whitewashed wall, also known as the "plastered brick wall", is a wall processed by plastering and whitewashing. Most of the whitewashed walls in palace buildings are iron red, while those in south China are white, and those in the north, greenish-gray.

花瓦顶是建造院墙墙帽常采用的方式。墙帽的种类有很多，用瓦砌成的常见的有蓑衣顶、眉子顶。有与屋顶相同的各式墙帽，还有花样很多的花瓦顶、花砖顶。其中花瓦顶就是对墙帽部分采用花瓦的做法，花瓦做法也用于园林的院墙、门楼、屋脊等处，有非常好的装饰作用。常见的花瓦图案有很多

The flower-tile top is a form of wall cap commonly adopted in constructing the courtyard walls. There are many types of wall caps. The wall caps made of tiles may be seen in walls with the straw raincoat top and the eyebrow-shape top, as well as the flower-tile top and the flower-brick top, which come in many various forms. The flower-tile top refers to the use of flowery tiles on the wall

• 花瓦顶
Flower-tile Roof

种，如官式做法沙锅套、十字花、锁链、竹节、长寿字、假叶子、银锭、辘轳钱等，还可以对这些基本做法进行组合使用。

墀头指硬山式建筑的山墙前后向外延伸，超出檐柱的部位。山墙是砌于房屋侧面的墙体，因硬山式建筑墙的上部与屋顶相接处为“山”字形而得名。墀头在建筑中处在比较显眼的位置，因此，其装饰备受重视。例如北京四合院中的墀头装饰，题材内容非常丰富，大多为有着吉祥寓意的图案。

cap. It is also seen in garden walls, gate towers, house ridges, etc. to achieve good decorative effects. There are many kinds of patterns for the flower tile, such as the official editions of the casserole cap, cruciate flower, chains, bamboo sections, longevity characters, fake leaves, silver ingot and windlass coins. These basic patterns can also be used in combinations.

Chitou refers to the portion of the gable of the flush-gable construction that extends beyond the eaves columns in the front or in the back. The mountain-shaped gables are used as the side walls for a house. The mountain gable derives its name from the Chinese character for mountain because it connects with the flush-gable roof of the house and forms a shape similar to the Chinese character of mountain (山). Since the *Chitou* is located at a rather obvious position of a building, it is therefore particularly featured as decorative. For example, the *Chitou* decorations of the Beijing quadrangle dwelling feature plenty of themes and most of them are patterns with auspicious significance.

房屋山墙

Mountain-shaped Gable of a House

墀头

Chitou

影壁在南方又叫“照壁”或“照墙”。其形态可以是一堵独立的墙体，也可以依附于其他墙体。被设置在大门之内的叫“内影壁”，被设置在大门之外的叫“外影壁”。

影壁由壁座、壁身、壁顶三部分组成，壁身部分往往被重点装饰，用彩画、泥塑、砖雕等进行装饰，内容涉及人物、花鸟、器物、文字等。而砖雕主要位于壁身中央，四角辅以装饰。

The screen wall or shadow wall, also known in the south as "lighting screen" or "lighting wall", is a type of wall that can either stand independently or be attached to other walls. When set within the main gate, it is called the "inner screen wall". When set outside the main gate, it is known as the "outer screen wall".

A screen wall is composed of three parts, namely the wall base, the wall body and the wall top. Decorations are focused on the wall body by means of coloring, clay sculpture and brick carving, with such contents as personages, flowers, birds, artifacts and Chinese characters, etc. Brick carving is primarily used on the central part of the wall body and also on the corners as auxiliary decorations.

上海玉佛寺影壁砖雕“万象更新”

The Brick Carving of "All is Renewed in Freshness" on the Screen Wall in the Jade Buddha Temple, Shanghai

• 平遥砖影壁上的麒麟砖雕
The Brick Carving of Kylin Patterns on the Screen Wall in Pingyao

• 砖影壁上大大的“福”字装饰
Large Chinese Character "福 (Meaning Luck)" Used as a Decoration for the Screen Wall

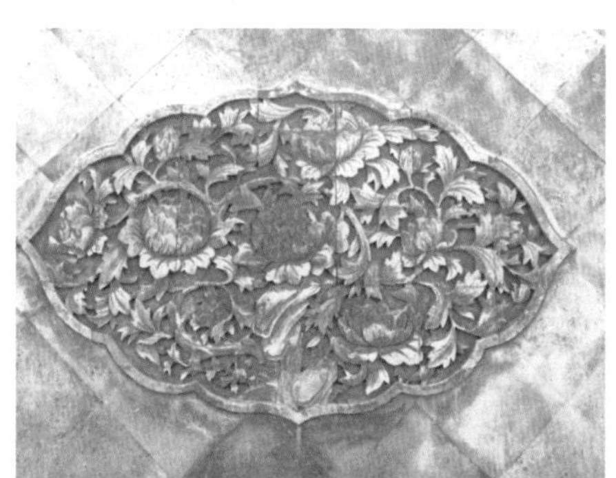

• 北京颐和园文昌园影壁上的宝相花砖雕
The Brick Carving of Composite Flowers on the Screen Wall in the Wenchang Garden in the Summer Palace, Beijing

• 北京颐和园仁寿门云龙纹砖雕
The Brick Carving of the Loong and Cloud Pattern on the Renshou Gate in the Summer Palace, Beijing

琉璃影壁为宫殿、皇家园林、礼制坛庙所用，壁顶多采用庑殿式或歇山式，壁座采用须弥座式，壁身饰以龙纹。龙纹有一龙、三龙、五龙、七龙、九龙之分。中国现存有三座大型的琉璃影壁“九龙壁”。第一座位于山西大同明代代王府门前，建于明永乐年间，壁面阔近46米，是三座中最大的一座。第二座位于北京故宫皇极门前，建

The glaze screen wall is used in palaces, imperial gardens and altars and temples as specified in the ritual system. The wall top usually comes in the hipped-roof style or the gable and hip roof style and the wall base is the Buddhist Sumeru base, while the wall body is where the decorative loong pattern rests. The loong pattern may feature one, three, five, seven

• **大同明代代王府九龙壁**

九条龙绚丽闪耀，昂首摆尾，弯曲盘绕，在海波上翻腾，犹如真龙再现，栩栩如生。该九龙壁面阔45.5米，高8.0米，厚2.2米。

Nine-loong Wall in Prince Dai's Mansion of the Ming Dynasty (1368-1644), Datong

The nine loongs are represented in glorious splendor. With their heads held high, flicking tails and coiled bodies, they plough through the waves of the sea and look realistic. The entire wall of nine loongs has a surface of 45.5 meters in width, 8.0 meters in height, and 2.2 meters in thickness.

于清代。第三座是位于北京北海公园北岸的琉璃九龙壁，是这三座中唯一的双面壁，做工也最为精美。

or nine loongs. There are currently three large-scale glaze nine-loong screen walls in China. One is located in front of the Ming Palace in Datong, Shanxi Province, which was built during the Yongle Period (1403-1424) of the Ming Dynasty (1368-1644). It has a surface width of nearly 46 meters and is the largest among the three. The second one, built in the Qing Dynasty (1616-1911), is located in front of the Gate of Imperial Supremacy in the Forbidden City in Beijing. The third, located on the northern shore of Beihai Park in Beijing, is the only two-sided wall among the three. Its workmanship is also the most beautiful.

- **北京故宫皇极门九龙壁**

皇极门是乾隆帝退位后的居所宁寿宫的第一道大门，地位非同一般。3500平方米的宫前广场南边，与皇极门相对的位置有一座面阔29.4米、高3.5米的五彩琉璃照壁，俗称“九龙壁”。壁下部为汉白玉石须弥座，上覆庑殿顶，壁面由270块彩塑琉璃砖组成。

Nine-loong Wall at the Gate of Imperial Supremacy in the Forbidden City, Beijing

The Gate of Imperial Supremacy is the very first gate of Palace of Tranquil Longevity, where Emperor Qianlong resided after his abdication. It enjoys, therefore, a very unusual status. On the south side of the Palace's 3,500-square-meter front square, a multicolored glaze screen wall, commonly known as the "Nine-loong Wall", is located just opposite to the Gate of Imperial Supremacy, measuring a surface width of 29.4 meters and a height of 3.5 meters. The lower part of the screen is the Buddhist Sumeru base made of white marble. With a hipped top, the screen's body surface was made of 270 colored sculptured glass bricks.

• **北京北海公园九龙壁**

北京北海公园北岸有一座琉璃九龙壁，原本在一座庙宇——大圆镜智宝殿山门前，该庙被毁，现仅存九龙壁。此九龙壁建于清乾隆二十一年(1756年)，壁高6.65米，厚1.42米，面阔25.86米，底座为青白玉石台基，上有绿琉璃须弥座。壁的双面各有用琉璃砖烧制的红、黄、蓝、白、青、绿、紫七色蟠龙9条。蟠龙戏珠于波涛和云雾之中，形态各异，栩栩如生。九龙壁之顶为庑殿顶，五条脊上共有龙32条，筒瓦、垄陲、斗拱下面的龙砖上都各有一条龙。如此算来，九龙壁上共有龙635条。整个影壁色彩绚丽，古朴而大方，是清代琉璃建筑中的杰作。

Nine-loong Wall in Beihai Park, Beijing

The Nine-loong Wall in Beihai Park, Beijing was originally located in front of the main gate of the Temple of Grand Wisdom of Round Mirror before the temple was destroyed. The Nine-loong Wall survived. Built in the 21st year (1756) during the reign of the Qianlong Emperor in the Qing Dynasty, this screen measures 6.65 meters in height, 1.42 meters in thickness and 25.86 meters in surface width. It is stationed on a greenish-white jade stylobate with a sumeru base made of green glaze. Both sides of the wall contain nine curled-up loongs in seven colors (red, yellow, blue, white, cyan, green and purple) made of fired glaze bricks. The lifelike loongs are depicted in different poses, playing with pearls in the clouds or in surf. The Nine-loong Wall has a hipped top and a total of 32 loongs on its five ridges. Each brick under the round tiles, border ridges and brackets has a loong on it. In other words, the wall has a total of 635 loongs. With gorgeous colors and primitive simplicity, this loong wall is truly an architecture masterpiece of glaze work from the Qing Dynasty.

九龙壁

龙壁是一种传统的建筑形式，九龙壁为最高等级。九龙壁是影壁的一种，通常建在帝后、王公居住或经常出入的宫殿、王府、寺院等建筑正门的对面。有砖雕、泥塑、彩绘、琉璃等多种形式。

Nine-loong Wall

The loong wall is a form of traditional architecture, and the Nine-loong wall is the highest form of it. The Nine-loong wall is a type of screen wall, usually built opposite to the front gate of the palace and estate where the emperor, empress, princes and dukes lived, or the temple that they frequently visited. It may be represented in many different forms such as brick carvings, clay sculptures, colored paintings, colored glaze and the alike.

• 清代黄木浮雕九龙壁

Nine-loong Wall with the Relief on the Yellow Rafter Wood in the Qing Dynasty (1616-1911)

木影壁立于宅院的二门之内，为以两对夹杆石夹起的一面木制的板墙，有顶。木影壁的中部是艺术装饰的重点区域，有的采用木雕纹样，有的采用吉祥文字。

The wooden screen wall, which stands between two gates in a courtyard, is a board wall clamped by two pairs of clip stones, with a roof on top. The middle part of its facade is the focus of decorative art in the form of either wood carving patterns or auspicious Chinese characters.

- 太极殿五福捧寿纹木影壁
  Wooden Screen Wall with the Traditional Five-bat Patterns and Longevous Pattern in Hall of the Supreme Principle

- 老舍故居七彩木影壁
  Seven-colored Wooden Screen Wall in Laoshe's Former Residence

> 门

门的作用不仅仅在于出入，还有隔断内外环境空间的作用，同时也是居住者身份的象征。门又被称为“门面”“门脸”，说明人们对门的重视。宫殿、宅院的大门为门楼建筑，一般也简称为“门”。门楼的造型、用料和做工代表着宫

> The Door

The role of a door is not simply to provide entry and exit, but also to separate the space of the internal and external environments. It also serves as a symbol of status of the occupants. As the access portals, doors are also known as "door appearance" or "door face", which expresses the emphasis people

婺源江湾萧江宗祠正门门楼

江西婺源江湾萧江宗祠为廊院式祠堂，有仪门、享堂、寝堂三进。正门门楼考究，门楼上装饰着精美的木雕图案。

Front Gate Tower of the Xiao and Jiang's Ancestral Temple in Jiangwan, Wuyuan

The Xiaojiang Ancestral Temple in Jiangwan, Wuyuan is an ancestral temple in the style of corridor yard. It contains three sections: the ceremonial gate, the offering hall and the resting hall. The gate tower of the main entrance is exquisitely wrought with the door beautifully decorated with wood carving patterns.

殿、宅院的社会等级。宫廷建筑、礼制坛庙、皇家陵墓的大门楼往往是一组建筑群，并不是单一的门楼；民间的祠堂建筑也是如此。门楣、门板、屋顶等构件通常被施以雕刻，运用圆雕、透雕等手法，雕刻出各种吉祥图案，以突出门楼的精巧和美观。门楼常以较为鲜艳的色彩进行重点装饰。门楼上的彩画都有不同的寓意，表达平安、幸福、子孙满堂等美好心愿。

垂花门的檐檩下不设立柱，而改做倒挂的莲花垂柱，造型美观，民居、府邸、寺观、园林大多采

• 山西王家大院垂花门
Festooned Door of the Wang's Courtyard in Shanxi Province

place on doors. The doorway structure of palaces and courtyards is known as gate towers, generally shortened as doors or gates. The shapes, materials and applied workmanship of a gate tower represent the social rank of the palace or courtyard. The large gate towers of palace buildings, ritual altars, temples and imperial mausoleums are often architectural groups instead of singular gate tower. This is also true with the architecture of ancestral shrines in civilian societies. The sense of sophistication and beauty of a gate tower will be increased by the sculpture work done in such parts as the lintels, door boards, roofs and other components of the gate tower. Through such techniques as circular carving and openwork carving, a variety of auspicious patterns are carved onto the gate towers to make it more delicate and good-looking. Gate towers are also highlighted by the decoration in bright colors. The colored paintings may carry various auspicious connotations, such as good wishes and prayers for peace, happiness, and productive household.

The festooned door is featured with the standing columns under the cross beam being replaced by two upside-down columns with lotus flower

用。古人所说的“大门不出，二门不迈”中的“二门”即垂花门。垂花门装饰性极强，各种装饰手段，如砖雕、木雕、石雕、油漆彩画均可使用，因此它在建筑、艺术方面的价值可谓极高。

垂花门外侧的梁头一般被雕成云头形状，叫“麻叶梁头”。麻叶梁头之下有一对倒悬的短柱，柱头向下，头部被雕成莲瓣、串珠、花萼云、石榴头等形状，酷似一对含苞待放的花蕾，这对短柱被称为“垂花柱”，垂花门的得名与这对垂柱有关。联结两垂柱的部件上也有很美的雕饰。所有的木构架上都有油漆彩画，以冷色为基调，配以五彩缤纷的小幅彩画。

垂花门柱间为雕刻精美的花枋，梁架上施以彩绘。门柱大体可分为中柱式和一殿一卷两种形式。中柱式即于面阔方向立柱，上安檐梁，以便承载上顶；一殿一卷式则表现为在中柱式垂花门后接出一个卷棚顶，以立柱支撑，形成复合建筑。遂初堂前的垂花门即为一殿一卷式。

洞门指在墙体上开设的不装门

patterns hung in midair. Attractive in appearance, festooned doors are used in civilian residences, official mansions, temples, shrines and courtyards. The "second door" mentioned in the ancient saying of "Not exiting the main door, nor stepping out the second door" refers to the festooned door. Festooned doors are highly decorative. Almost all the decorative means, such as brick carvings, wood carving, stone carving and colored painting are utilized on festooned doors. Therefore, it has rather high value in architecture as well as in art.

The head of the outside beam of the festooned door is usually carved into the shape of a cloud, called the "Ma-leaves beam head (*Maye Liangtou*)". Underneath the Ma-leaves beam head are a pair of short upside-down columns with the heads of the columns facing downward and carved with various decorative shapes, such as lotus petals, beads, calyx-like cloud and pomegranates looking like a pair of budding flowers. This pair of short columns is called the "hanging flower columns", which explains the name of the festooned door. The components connecting the two vertical columns are also beautifully carved. All of the wooden frames are painted with colored patterns

• **宁寿宫花园遂初堂垂花门**

在宁寿宫西路的乾隆花园中的形制完整的垂花门，是为乾隆皇帝的休闲生活营造园林气氛而建。

Festooned Door of Hall of Wish Fulfillment in the Garden of the Palace of Tranquil Longevity

The fully preserved structure of the festooned door of Qianlong Garden located on the West Ningshougong Road was originally constructed to create a landscaped garden for Emperor Qianlong.

扇的门洞，可以砌成各种形状，如圆形、方形、瓶子形、葫芦形、海藻形、秋叶形、桃形等。洞门上端雕刻有海棠纹、回纹、云纹、角花等纹样。洞门多用于园林中的园墙之上，通过不同形状的洞门欣赏景物，可以获得不同的观景感受。

based on cool colors and sometimes supplemented with small fragments painted in riotous colors.

In between the doorposts of the festooned door are the square beams carved with flower patterns, with the beam frames color-painted. The columns can be generally arranged in the central-post style and the one-hall-one-roll forms. In the central-post style, the posts are set up in the direction of the *Miankuo*, with beams placed on them to support the top. The one-hall-one-roll form adds an additional round-ridge roof to the central-post styled festooned door, with the round-ridge roof supported by a standing post, thus forming a composite construction. The festooned door in front of Hall of Wish Fulfillment was completed with a one-hall-one-roll structure.

A caved door refers to an opening to a wall that does not have any door leaf installed. The opening can be in various shapes, such as a circle, square, gourd, seaweed, autumn leaf and peach. The top portion of the caved door is generally sculpted with patterns of begonia, fretwork, cloud or angular flower. Caved doors are typically utilized for garden walls in courtyards. Looking through the variously-shaped caved door allows viewers to experience different feelings from the landscape.

• 谐趣园瓶形洞门

Bottle-shaped Caved Door at the Garden of Harmonious Interests

• 海棠洞门

Begonia-shaped Caved Door

隔扇门也称“格扇门”，是被安装在金柱或檐柱间的门，也是中国传统建筑最常用的门扇形式。隔扇门一般用于明间的装饰，整排使用，通常为四扇、六扇或八扇。格扇门的上部为隔心，由花样的棂格拼成，可透光；下部为裙板，不透光，可以有木刻装饰。还有的被做成落地的长窗，即隔扇全用隔心，而不用裙板。大片的空心窗格花纹可产生极富变化的韵律感，裙板上丰富多彩的雕刻增强了房屋的装饰效果。

The partition door, also known as "fan door", is a door installed between hypostyle columns or eave columns. It is the most commonly used door in traditional Chinese architecture. A partition door is generally used as decoration for the main room in the structural center of the house. It is typically used in a row of four, six or eight. The upper part of the door, called partition center, is made up of patterned window grids, which are pervious to light. The lower part, skirt board, impervious to light, can be decorated with wood carvings. In some cases, the whole piece is made pervious to light down to the ground like a long window, that is, the partition door is all partition centers with no skirt boards. A large piece of hollow patterned pane can produce a rich rhythm full of change and the variety of carvings on the skirt boards can enhance the decorative facade of the house.

安徽绩溪胡氏宗祠寝宫隔扇门

胡氏宗祠寝宫与两庑的隔扇门形制相同，门上疏密有致的隔心用于寝宫的采光。门的裙板上的图案为花瓶。隔扇门下是高高的门槛。

Partition Door at the Mausoleum of Hu's Ancestral Temple in Jixi, Anhui Province

The partition door of the resting chamber in Hu's Ancestral Temple and those of its two side rooms are identical in form. The orderly arranged partition centers on the door provide the resting chamber with natural light. The skirt boards are decorated with vase patterns. A high threshold is used below the partition door.

门匾

门匾一般被挂在门的上方、屋檐的下方，具有一定的装饰作用。古人用门匾表达经义、感情或建筑物的名称和性质。横着的叫“匾”，竖着的叫“额”。匾额中有的形如书卷，被称为“手卷形匾”，形式较为优美，多用于园林的小型建筑中，如亭榭、书斋等处的屋檐下或室内。匾额中形如册页者叫“册页形匾”，形式也很优美，多用于园林建筑。叶形匾是一枚树叶之形的匾，这种活泼形制的匾多见于园林之中。荷叶匾即卷荷叶之形的匾，是一种少见的形制的匾。虚白匾即以透刻花纹为底的匾，没有明显的底板，也叫“石光额”，大多用于园林中，是一种富有装饰性的匾。

• 手卷形匾
Scroll Plaque

The Door Plaque

A door plaque is generally hung above the door just under the roof. It has a certain decorative effect. Door plaques were used by ancient people to express comments, feelings, or the name and nature of their buildings. The horizontal door plaque is called "*Bian*", and the vertical one is called "*E*". Some of the door plaques come in the shape of scrolls with graceful forms, called the "scroll plaque", which are mostly placed under the eave or inside small garden structures, such as pavilions and studies. There are also some plaques in the shape of book pages called "page-shaped plaque" which are elegant in form and are often used in courtyard architectures. Leaf-shaped plaque is a vivacious design frequently seen in landscaped gardens. The lotus plaque is in the shape of a curled lotus leaf, which is a rare design. The white-blank plaque usually features hollowed-out patterns at its base with no apparent background plate, hence it also called "stone-light plaque". It is the kind of plaques rich in decorations, mostly used in landscaped gardens.

虚白匾

White-blank Plaque

乾清宫“正大光明”匾

"Fair and Square" Plaque at the Palace of Heavenly Purity

北京故宫益寿斋蝙蝠如意匾

"Bats and *Ruyi*" Plaque of Lodge of Longevity in the Forbidden City, Beijing

北京故宫怡性轩册页形匾

Page-shaped Plaque of Yixing Pavilion in the Forbidden City, Beijing

平遥乔家大院“会芳”荷叶匾

匾额上有“会芳”二字，匾为绿色，字体被贴金，十分典雅。

Lotus-shaped Plaque with the Chinese Characters of "*Huifang*" at the Qiao's Courtyard in Pingyao

The plaque is green with gold-colored Chinese characters "*Huifang* (Collection of aroma)" in an elegant way.

门前装饰

在古代，不仅门本身的装饰极为重要，门前的装饰也很受重视，等级越高的建筑，其门前装饰的等级也相应越高。门前的装饰主要包括门墩、滚墩石、拴马桩、上马石、石狮、铜铸饰品等。

门墩又称“门鼓石”“抱鼓石”“门枕石”，常见于中国传统民居四合院的大门底部，起到支撑门框、门轴的作用。整体称“门枕石”，门外部分称“门墩”。门墩上面一般雕刻有精美的吉祥纹样。人们借助这些图案表达了希望长寿、富贵、生活美满、家族兴旺的美好心愿。

滚墩石常用于垂花门和影壁两侧，故又名“垂花门滚墩石”和“影壁滚墩石”。滚墩石的形状与门墩类似，不同之处在于滚墩石是对称的，从造型上看，滚墩石表现为一对门墩对接在一起的样态。

古代民居门前的拴马桩被埋立在大门前两侧的墙根下，或者并排立在院子的里门旁，柱头上雕刻着狮子，又称“狮子柱”，也有被雕成猴头、“仙人”、桃尖等形状的柱头。拴马桩是北方独有的民间石刻艺术品，不仅可以用来拴马，还可以保护住宅墙面，使之不受碰撞。而且它与门前的石狮一样，既有装点建筑、展示富有的作用，同时还被赋予了保祐全家平安的意义。

上马石是便于古人上马的石台，以细青石或汉白玉制成，有的上马石被做成阶梯式，设置在大门前端，其上雕刻有各式各样的花纹、历史故事图等。旧时北方的

• **山西王家大院台阶边的抱鼓石**

抱鼓石上的《蟾宫折桂图》展现的是一个身穿状元服的人骑着马，身后有人举着华盖。

Drum-holding Stone by the Steps at the Wang's Courtyard in Shanxi Province

The picture of *Plucking the Laurel Branch from Toad Palace* (a Chinese proverb implying success in the imperial examination) on the drum-holding stone shows an exam champion riding a horse, with people holding up a canopy behind him.

山西王家大院门前的抱鼓石

雕刻的刘海戏金蟾，造型古朴、大方。

Drum-holding Stone in Front of the Wang's Courtyard in Shanxi Province

Carved with the motif of Liu Hai playing with the golden toad, the stone is stylized with primitive simplicity and open-mindedness.

太原晋祠门口的抱鼓石

Drum-holding Stone in Front of Jinci Temple, Taiyuan

府第和大四合院、大会馆的门前都被放置上马石。住宅门前有没有上马石也是划分宅第等级的一个标准。

石狮在宋代以前是帝王陵墓石雕，有显示帝王尊严、权威的作用。元代以后，石狮的使用范围扩大，宫殿、苑囿、官署等处均可使用，它成为门前建筑装饰的重要组成部分。石狮又是官阶等级的象征。古时规定：一品官员门前的石狮头上的鬃毛卷为13个，称为“十三太保”，二品官员门前的石狮头上的鬃毛卷为12个，依次类推，七品官员门前的石狮头上的鬃毛卷为7个，但七品以下官员门前不准摆放石狮。

门钉最早用于板门之上，其主要作用是加固大门，后来逐渐演变成了装饰和等级的象征。明代时对门钉的数量无规定，一般是纵、横三至五路，每路三至五

• 滚墩石

Rolling Mound Stone

• 平遥建筑门前的拴马桩

Horse Hitching Pole in Front of the Door of a Building in Pingyao

• 山西王家大院门前的上马石
Horse-Mounting Stone in Front of the Wang's Courtyard in Shanxi Province

枚。清代时，门钉的数量与建筑物的等级有关，如皇宫大门纵、横各九路，每扇门81枚门钉，象征帝王最高的地位。五品以下官府的大门不准设门钉。

门簪被安在街门的门楣之上，有两个或四个，有正方形、长方形、菱形、六角形、八角形等样式，正面以雕刻、描绘等手法饰以花纹、文字等图案。门簪在汉代就已出现，初时起到固定门扇的作用，后来其装饰性大于实用性。门簪的图案以四季花卉较为常见，四枚分别雕以春兰、夏荷、秋菊、冬梅，图案间还常见“吉祥如意”“福禄寿德”“天下太平”等字样。门簪为两枚时，则常雕以“吉祥”“如意”等字样。

辅首又称“门辅”，是被安装在大门上，用于衔门环的一种底座，也是传统建筑大门上的一种装饰。汉代时，大门上就已经出现辅首，至今已有将近2000年的历史。民用的辅首一般被安装在最主要的大门上，造型简单，多为圆形，铁制或铜制。帝王宫殿大门上的辅首等级较高，铜质镏金，形象多为虎、螭、龟、蛇，有吉祥之寓意。宫殿大门辅首也有采用瞪目、张口的狮头造型的，既有守门之意，又显示了皇家建筑的威严和庄重。

Decorations in Front of Doors

In ancient times, not only were the decorated doors extremely important, but the decorations in front of the door were also highly regarded. The higher grade the architecture, the higher the class of the decorations in front of the door. The decorations used in front of the door include the door mound rolling mound stone, horse hitching pole, horse-mounting stone, stone lions, cast bronze ornaments and so on.

The door mound, also known as "door drum stone" "drum-holding stone" and "door pillow stone", is commonly seen on the bottom of the main entrance to the traditional Chinese residential quadrangle dwelling. It is a stone component used to support the door frame and the

- **天安门前的石狮**

天安门前的石狮是北方石狮的代表，庄重而威严，勇猛而大气。

Stone Lions in Front of Tian'anmen

The stone lions in front of Tian'anmen are the typical northern stone lions with a solemn, dignified and majestic air.

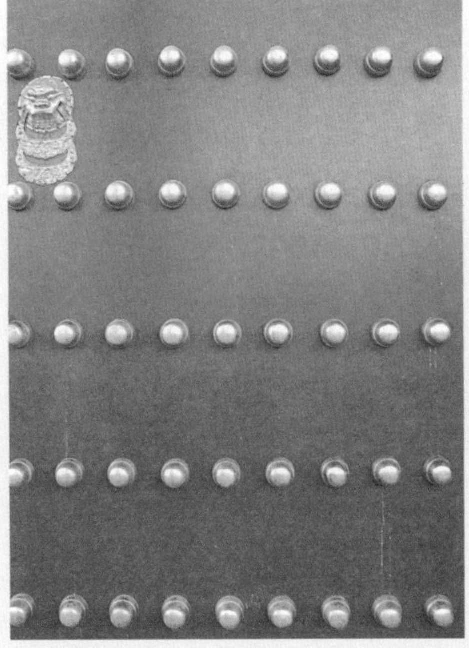

- **午门的门钉**

门钉一般用于实榻门（由原木板拼接而成的实心大门，在大门中等级最高，体量最大）正面，起加固大门、防止门板松动的作用，更是一种皇家威仪的象征。

Door Nails on the Meridian Gate

Door nails are generally used on the front surface of *Shi Ta Men* (doors made of solid wood boards spliced up, which ranked the highest in class and the largest in body) to reinforce the door and prevent the loosening of the boards. Moreover, it was a symbol of imperial dignity.

door shaft. It is generally called the door pillow stone as a whole, while the portion outside of the door is called the door mound. The surface of the door mound is typically carved with delicate auspicious patterns. People make use of these patterns to express their hope for longevity, wealthy, happy life and family prosperity.

The rolling mound stone, commonly used on both sides of the festooned door or the screen wall, is also known as the "festooned-door rolling mound stone" or "screen-wall rolling mound stone". The shape of the rolling mound stone is similar to that of the door mound. The difference lies in the symmetry of the rolling mound stone. In terms of its design, a rolling mound stone looks like two drum stones put together.

The horse hitching pole in front of ancient people's houses was buried into the ground by both sides of the lower portion of the wall in front of the main entrance, or set up in a row by the side of the yard door. If the top of the horse hitching pole is carved with a lion, it is also called a "lion pole". It can also be carved with a monkey head, an immortal or the tip of a peach. The horse hitching pole is a unique folk art of stone-carving in the north of China. Not only can it be used to tie horses, but also to protect the house wall from impact. Like the stone lions in front of doors, it decorates the architecture to show the wealth of the residence and is also endowed with the significance of protecting the house.

The horse-mounting stone is a stepping stone used by ancient people to mount a horse with more ease. It is made of fine blue stone or white marble stone. Some are made into the shape of a ladder and placed in front of the door. A variety of patterns and pictures of historical stories are carved onto the horse hitching stone. The old northern mansions, large quadrangle dwellings and big meeting halls used to have horse hitching stones placed on both sides of the doors. The presence

• 北京的四合院大门：门簪上有“吉祥”二字
Main Gate of the Quadrangle Dwelling in Beijing: the Door Cylinders Have the Characters "*Jixiang* (Good Luck)" on Them

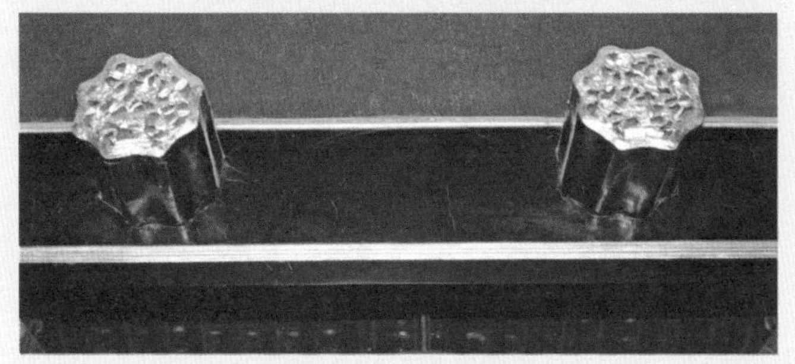

• 北京恭王府垂花门的门簪
Door Cylinder on the Festooned Door of Prince Gong's Mansion, Beijing

of a horse hitching stone also served as a grading criterion for the classification of the residence.

Prior to the Song Dynasty (960-1270), stone lions were used in the imperial mausoleums to display the dignity and sovereignty of the emperor. After the Yuan Dynasty (1206-1368), the use of stone lions was extended to palaces, parks, gardens and government offices. It became an important architectural decoration in front of the door. Stone lions were also a symbol of official ranks and grades. Ancient regulations required that the stone lions in front of the first-grade officials' residence should each have a total of 13 curled whiskers and were thus called the "thirteen heroic ones". The stone lions in front of the second-grade official's residence should each have a total of 12 curled whiskers. It goes on like this. The stone lions in front of the seventh-grade official's residence had a total of 7 curled whiskers, but officials lower than the seventh grade were not allowed to place stone lions in front of their residences.

Door nails were first used on large doors as reinforcement but they gradually became a decoration and a symbol of class. During the Ming Dynasty (1368-1644), there were no specific regulations on the number of door nails on a single door. Generally, it was three to five rows both horizontally and vertically, with three to five nails on each row. In the Qing Dynasty (1616-1911), the number of doornails was related to the class of the architecture. For example, the nine rows were set both vertically and horizontally in front of the Imperial Palace with 81 nails in all given to each door to represent the highest status of the emperor. However, officials below the fifth grade were not allowed to put nails into their doors.

The door cylinder is secured on the central threshold of street doors, used in twos or fours, and made in square, rectangular, diamond, hexagonal or octagonal shapes. Its front side is decorated with carvings, colored paintings or else in patterns of flowers, Chinese characters, etc. The door cylinder was already in existence in the Han Dynasty (206 B.C.-220 A.D.). It was initially used to fix the door leaf while, later on, its role as a decoration soon surpassed its practicality. Door cylinders are commonly patterned with seasonal flowers. When they come in fours, they may each feature orchid flowers for the spring, lotus flowers for the summer, chrysanthemums for the autumn and plum blossoms for the winter. Frequent patterns include phrases such as "good luck and happiness" "happiness, good fortune, longevity and virtue", "peace and harmony everywhere", and so on. In the case of pairs, words like "good luck" "*Ruyi* (everything goes well)", and others are often carved.

The door knocker ("head support" literally), also known as "door auxiliaries", is a base design installed onto the door to hold the knocking ring. It is also a type of door decoration used in traditional architecture. The door knocker has been used on doors for more than 2,000 years since as early as the Han Dynasty. Door knockers were used by civilians primarily on major doors. They were simple, mostly round in shape, and made of iron or copper. The door knockers used on the gate doors of the imperial palaces are of higher class. They are made of copper gilt with gold and generally shaped in the image of a tiger, hornless loong, turtle or snake to symbolize the zodiac guardians that implied good luck. There are also door knockers in the shape of a staring and open-mouthed lion to represent the gatekeeper and signify the majesty and solemnity of imperial buildings.

北京故宫景运门上的辅首

故宫大门的辅首为椒图形象，眼睛、鼻子及眉毛上的装饰被刻画得很细腻，图案使得辅首大气十足。

Door Knockers on the Door of Jingyun Gate in the Forbidden City, Beijing

The door knocker on the door of Jingyun Gate in the Forbidden City is delicately carved with the decorative image of *Jiaotu* (one of the nine mythical sons of a loong) with detailed and vivid depictions of the eyes, nose and eyebrows, giving it a majestic air.

门神

Door Gods

门神是在民间人们认为守卫门户的“神灵”。旧时人们都将其“神像”贴于门上，用以保祐家宅平安，祈求成功、吉祥等。门神是民间最受人们欢迎的一类“保护神”。门神画是旧时农历新年贴于门上的一种画，春节时可将宅院装点一新。

The door god denotes a belief in a divine door-guardian shared by the general public. In ancient times, people would paste pictures of the door god on doors to protect the safety of the house, and pray for success and good luck. It is one of the most popular gods in the world of the general public. Paintings of door gods were also posted on doors during the Lunar New Year Festival in ancient times as a kind of decoration in spring.

门神秦琼、尉迟恭

秦琼和尉迟恭都是唐代初期著名的大将。相传唐太宗在晚上睡觉时常常听到卧房外边有呼叫，抛砖、掷瓦之声，故而夜不成眠。于是他便与众臣商议，令秦琼和尉迟恭二人把守宫门，夜晚果然平安无事。唐太宗感念二人辛劳，于是将二人的画像贴在门上，照样管用。后来，民间也开始流行贴二人画像于门上。

The Door Gods Qin Qiong and Yuchi Gong

Qin Qiong and Yuchi Gong were both well-known generals from the early Tang Dynasty. According to legend, Emperor Taizong of the Tang Dynasty often heard calling, and throwing bricks and tiles outside his chamber while he was trying to sleep, which caused him could not sleep well. After consulting the officials in his court, he ordered Qin Qiong and Yuchi Gong to guard the gate to the imperial harem. Sure enough, the nights went by safe and sound disapperd ever after. Later, as a thankful repayment for the labor of the two generals, Emperor Tang Taizhong chose to paste the portraits of the two generals on the door and it was just as effective. Subsequently, pasting the portraits of the two generals also became popular among ordinary people.

门神神荼、郁垒

神荼、郁垒是传说中可以捉恶人的“神”。如果有恶人想祸害他人，神荼、郁垒便会将他们抓住。古人在辞旧迎新之际，在桃木板上分别写“神荼”“郁垒”二神的名字，或者用纸画上二神的图像，悬挂、嵌缀或者张贴于门首，有祈福之意。

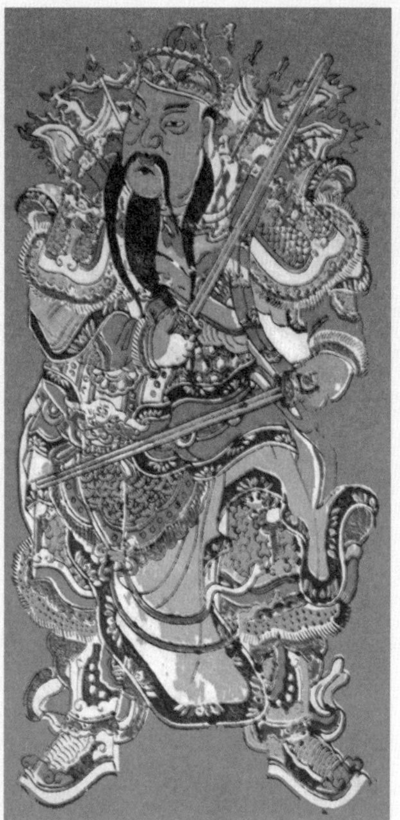

漳州年画《门神》（明代）

右为尉迟恭，左为秦琼。尉迟恭纯朴而忠厚，勇武善战，一生戎马，屡立战功。秦琼勇猛而彪悍，每战必先，常于千军万马之中取敌将首级。

Zhangzhou New Year Pictures Representing Door Gods (Ming Dynasty, 1368-1644)

The general on the right is Yuchi Gong, and the other on the left is Qin Qiong. A simple and honest man, Yuchi had proved his gallantry and superb fighting skills in numerous battles throughout his life and rendered illustrious military service. Qin, brave and fierce, led the charge in every battle and often dashed into the crowd of the enemy and chopped off their commander's head.

The Door Gods Shen Tu and Yu Lei

Legend has it that Shen Tu and Yu Lei were the gods who capture evil people. If malicious people tried to disturb others, Shen Tu and Yu Lei would capture them. During the Chinese New Year when people were bidding farewell to the old time and welcoming the new year in ancient times, people would write the names of "Shen Tu" and "Yu Lei" on mahogany boards separately, or draw the images of the two gods on paper, and then either hang, embed, or post them on the door as a prayer to wish for a blessing and to fend off evil.

门神哼哈二将

哼哈二将为守护寺庙的两位“神将”，形象威武而凶猛。一名“神将”叫郑伦，能鼻出白气制敌；一名“神将”叫陈奇，能口哈黄气擒将。

The Door Gods General Heng and General Ha

General Heng and General Ha are the guardian gods of temples who had a mighty and ferocious appearance. General Heng's real name is Zheng Lun, who can release white gas through his nostril to subdue the enemy, and General Ha, real name being Chen Qi, can release yellow gas from his mouth to capture the enemy.

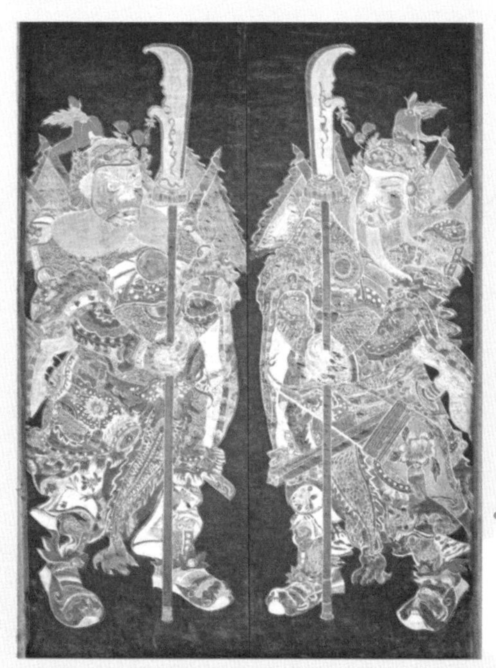

• 安徽绩溪龙川胡氏宗祠门上的神荼和郁垒

Images of Shen Tu and Yu Lei on Hu's Ancestral Temple at Longchuan Village in Jixi, Anhui Province

> 窗

窗是被安装在建筑物上用来采光、换气的构件。传统建筑中的窗分为两大类：一类是木棂格窗，在工艺上属于木作装饰；一类叫

• 灯笼框窗棂格

Grid Pattern Window of Lantern-frame Patterns

> The Window

The Window is installed in architecture to allow in lighting and ventilation. In traditional architecture, windows are divided into two categories. One is the wood grid-patterned window, which is considered decorative woodwork in industrial art. The other is *You* (windows between the room and the hall), an opening in the wall that is considered part of the masonry.

The wood grid-patterned window was the most extensively used window in ancient times. Since there was no glass in ancient China, the only way to solve the problem of indoor lighting and maintaining warmth was to glue paper onto windows. To ensure enough strength in window paper, the gridded-patterned window was utilized. Aside from being practical, grid-patterned windows also

“牖”，在墙上开窗，在工艺上属于瓦作。

木棂格窗在古代是被广泛使用的窗户类型，中国古代没有玻璃，解决室内采光和保暖问题的唯一办法是在窗上糊纸。为确保窗户纸具有一定的强度，故普遍采用棂格窗。除了实用，棂格窗上面雕刻的不同的纹样还可起到装饰作用。

镶嵌在隔扇、槛窗中的棂格以攒斗法拼成，式样很多，有龟背锦

played a decorative role when engraved with various types of patterns.

The grid pattern set in the partition doors and the caged-in windows is built by using *Cuan* (a method of assembling small vertical and horizontal wood pieces into geometrical patterns through mortise and tenon joints) and *Dou* (a method of assembling small sculptured pieces into patterns) into various styles, such as brocade grids in turtle-shell shapes, lantern-frame grids, ice-crack grids,

• 步步锦窗棂格

Window of *Bubujin* (Steps of Prospect) Rectangular Grids

• 古钱菱花棂格

Grids of Ancient Coin and Four-rhombus Patterns

支摘窗

Removable Window

北京故宫体元殿的支摘窗

支摘窗上部可依天气变化以纱、纸糊饰；下部安装有玻璃，以利室内采光。

Removable Window of the Hall of All-encompassing Universe in the Forbidden City, Beijing

The upper part of the removable window can either be pasted with gauze or paper according to the weather, and the lower part is usually installed with glass to allow in light.

格、灯笼框格、冰裂纹格、玻璃屉格、套方格、拐子锦格、盘长格、方眼格、豆腐块格、如意锦格、卧蚕锦格、四菱花格等。窗棂格花纹有美好之寓意，以象征的手法表现了人们对平安、幸福、欢乐、美满生活的向往，是中国门窗装饰构件中最精彩的部分。

glass-tray grids, square-set grids, brocade grids in cane shapes, Buddhist lucky pattern grids, square-eye grids, tofu-block grids, brocade grids in *Ruyi* shapes, brocade grids of sleeping silkworm, four-rhombus grids, and so on. The grid pattern of windows bears beneficial implications in representing in a symbolic way people's longing for peace, happiness, joyfulness and a fulfilled life. It is the most exciting part of the Chinese outdoor decoration of doors and windows.

北京北海公园琼岛半月城智珠殿槛窗

槛窗被安在槛墙之上、柱子之间，上下有转轴，便于开启和关闭。槛窗的设置有利于室内的透气、采光，增强建筑的通透性。槛窗的形式庄重，一般用于厅堂、园林等重要的建筑中。

Caged-in Windows at Zhizhu Palace Located in Banyue City of the Jade Istet in Beihai Park, Beijing

The caged-in or framed window is installed above the caged-in wall and in between pillars. With shafts at the top and the bottom, it can be easily opened or closed. The caged-in window is designed to provide indoor lighting and ventilation and to increase the permeability of a building. The dignified style of the caged-in window is generally expressed in such important structures as halls and courtyards.

冰裂纹窗棂格

Grid Pattern Window of Ice-crack Patterns

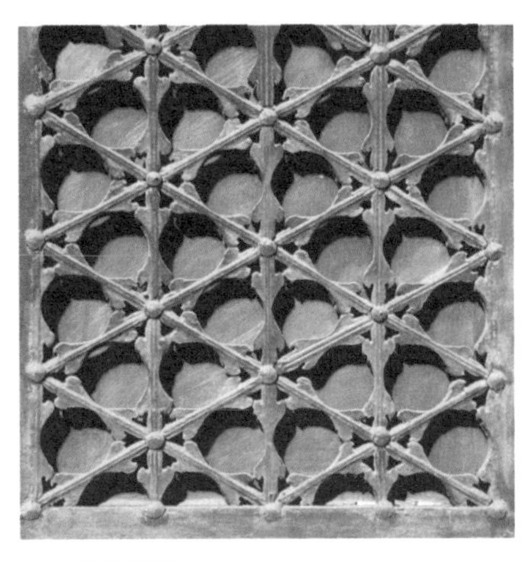

四菱花棂格

Four-rhombus Pattern Grids

承启楼内直棂窗

直棂窗是一种古代常用的窗户类型，造型比较简单。圆形土楼的窗户一般都是直棂窗，这样的窗户通风性极好。而且，圆形土楼窗前一般是通道、廊道或过道，这又增强了土楼的通风性能。

Vertical-grid Window of Chengqi Tower

The vertical-grid window (a window with patterns of vertical grids) is a type of window commonly used in ancient times. It is relatively simple in design. The windows of round clay buildings are generally vertical-grid windows. They provide excellent ventilation. Furthermore, since the windows of round clay buildings are generally facing passages, corridors and hallways, they can further reinforce the ventilation performance of the buildings.

漏窗俗称“花墙头”“漏花窗”“什锦窗”，多用在住宅、园林当中。窗框的制作材料有石、砖、木三种，通常的做法是把砖瓦件镶嵌在墙面上，构成各种精巧的图案。漏窗装饰始于明代中期，《园冶》一书中列举了十几种精巧、细致的漏窗，常见的形状有方形、圆形、六角形、八角形、扇形、菱形、花形、叶形等，窗内纹

The perforated window, commonly known as "flowery wall top" "perforated-pattern window" or "miscellaneous window", is generally used in residential houses as well as courtyards. The materials used for the frame of perforated windows include stones, bricks and wood. It is usually formed by inlaying bricks or tiles in the opening on the surface of the wall to create a variety of delicate patterns. The decoration of perforated windows began during the middle of the Ming Dynasty. In the book titled *Yuanye* (garden cultivating), more than a dozen exquisitely detailed perforated windows are cited. The most commonly seen shapes of perforated windows are squares, circles, hexagons, octagons, fans, rhombus, flowers and leaves. The inlaid patterns inside the perforated window include connected

- **苏州怡园碧梧栖凤馆铜钱纹花窗**

窗户借用铜钱的形式，不过将钱纹的圆形处理成了八角形。

Coin-patterned Perforated Window of the Hall of Phoenix Perching on the Green Tree in Yiyuan Garden, Suzhou

The window utilized the shape of a coin and changed the circular shape of the coin pattern to an octagonal one.

样有连钱、叠锭、鱼鳞、宫式、夔式、竹节、菱花、海棠等。各式各样的精美图案，虽是为装点风景而设计，但本身也是一种风景。

coins, stacked ingots, fish scales, palace-style, *Kui* (one-legged mythical monster)-style, bamboo joints, rhombus, begonia, and so on. The wide range of exquisite designs of perforated windows is meant to decorate the landscape, while itself is a scene too.

“万象更新”漏窗

一头憨态可掬的大象，身上的披巾上刻有一盆万年青，寓意为“万象更新”。大象周围饰有飘带，周围的棂格中还刻有葫芦纹和飘带纹 。

Perforated Window with the Pattern of "All is Renewed in Freshness"

An elephant in a drape inscribed with a pot of evergreen with the logo "All Is Renewed in Freshness". The elephant is surrounded by decorative ribbons and the surrounding grid structure is carved with the patterns of gourds and ribbon.

狮子林琴棋书画花窗

琴棋书画图案与窗巧妙地结合在一起，镂空雕刻的图案造型别致。

Perforated Window with the Pattern of Zither, Chess, Calligraphy and Painting in the Lion Grove Garden

The images of lute, chess, calligraphy and painting are cleverly integrated into the window, with the hollowed-out patterns in delicate styles.

> 斗拱

斗拱是中国传统建筑特有的构件，被置于柱头和额枋、屋顶之间，其作用是加大房檐，传递屋顶负荷，减少跨度。中国重要的传统建筑如宫殿、坛庙、寺观，以及大型楼台、亭阁等都设有斗拱。斗拱

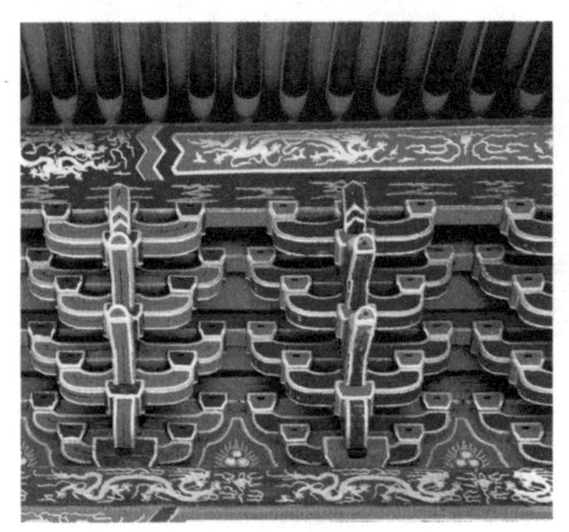

• 北京先农坛太岁殿镏金斗拱
Gold-gilded Brackets Used in the Taisui Hall of the Temple of Agriculture, Beijing

> The Wood Bracket

The wood bracket is a unique component used in traditional Chinese architectures. Placed between top of columns and architraves and the roofing, it can extend the eave, distribute the load of the rooftops and reduce the horizontal span. Such important Chinese traditional architecture as palaces, altars, temples, towers and pavilions make extensive use of wood brackets. The more layers of structure the bracket extends, the higher class the architecture is. Prior to the Ming Dynasty (1368-1644), the wood bracket used to be the structural components that bore load and acted as decoration at the same time. After the Ming Dynasty, the load-bearing function of brackets began to weaken, while its role as a decoration began to materialize. By the time of the Qing Dynasty (1616-

的层数越多，建筑的等级就越高。明代以前，斗拱是有承重作用的构件，同时也有装饰作用。明代以后，斗拱的承重作用逐渐减弱，装饰性增强，至清代时，已基本成为装饰物件。斗拱构造精巧，造型美观，如盆景，似花篮，而且表面装饰有彩绘图案，十分优美。

1911), wood brackets were essentially decorative items. With compact structure and delicate form, wood brackets are attractive in appearance like a bonsai or a flower basket, with the surface decorated with beautiful painted designs.

• 五踩斗拱
Five-step Bracket

> 柱

柱是房屋大型木结构中的承重构件，被竖立安装在台基的柱顶石上。柱子一般为圆柱体，也有四方、海棠式、六方、八方等异形柱，但用途都相同。柱子下端被插入柱础中，上端与枋木相连，形成稳定的结构。柱子因安装位置不同而有专门的名称，如檐柱、金柱等。

中国古建筑的木构架结构是由大木作（主要结构部分）开始施工的。台基建好后，就要安装立柱。柱子是承载屋顶的主要构件，柱子的高矮决定了建筑物房间的高度，柱子的位置决定了房屋开间和进深的大小。每根柱子的下端都有柱顶石，柱子的上端以榫卯结构与枋木和随梁枋相连。柱子多以油漆、彩绘的方式进行装饰。例如宫殿、寺

> The Pillar

The pillar is a load-bearing component of a building's large wooden structure, generally erected above the column base stone of the stylobate. Pillars are typically cylindrical in shape but may also be square, hexagonal, octagonal or begonia-shaped. Whatever shape it takes, its function remains the same. The bottom of the pillar is inserted into the column base and the top is connected to the square horizontal wood on top of pillars to constitute a stable structure. Pillars installed in different positions are differently named, such as the eave columns and the in-house columns.

The Chinese ancient architecture were built from structural carpentry. Once the stylobate is constructed, pillars are installed and serve as the main components to support the roof. The

• 被漆成红色的柱子
The Pillars of the Painted Red

庙、民居的柱子常被漆成红色，给人以喜庆、庄重之感；皇宫的柱子则被漆成金色，以显尊贵；园林亭廊的柱子多被漆成绿色，以呼应周围景色；等等。

length of the pillars determines the height of the rooms in the structure, while the positions of the pillars dictate the size and depth of the rooms. Every pillar has a pedestal boulder under its lower end, and its upper end is connected to the square horizontal wood on top of pillars and the horizontal beam accompanying the roof beam through tenons. Pillars are often decorated with oil or colored painting. For instance, pillars in palaces, temples and houses are generally painted red to express both festivity and dignity; pillars of the Imperial Palace are painted in gold as symbolic of the highest rank and those in courtyards, corridors and pavilions are typically painted green to echo the surrounding scenery; etc.

楹联

楹联指挂在楹（厅堂前面的柱子）上的对联。一般以木板、竹片做成弧形，其上雕以对联文字，之后再敷以红、绿、蓝、金、白等色，加以精致装潢，悬挂于楹柱上。楹联是文学和书法的结合体，符合中国人的审美习惯，成为传统建筑常用的装饰手段，在园林中有较为广泛的应用，既具有点景的作用，也是园林欣赏的重要内容。另外，还有一些形制特殊的楹联，更具装饰美感。

Column Couplets

Column couplets refer to the couplets that are written on scrolls and hung on the pillars in front of a hall. Column couplets are typically made of wooden or bamboo arc-shaped boards, with written couplets carved and colored in red, green, blue, gold or white on the surface of the boards. They hung on the pillars in front of a hall after being well decorated. The column couplets are a combination of literature and calligraphy with the aesthetic custom of Chinese people. It is a decorative method frequently used in traditional architecture and widely seen in courtyards. It is a means to decorate the landscape scene, as well as an important content in the appreciation of a garden. In addition, there are also some specially-shaped column couplets which claim more decorative beauty.

- **蕉叶联**

被做成芭蕉叶形状的对联叫“蕉叶联”。这种形状优美、别致的对联多见于南方私家园林中，常被挂于轩、榭、亭、阁的角落处。

Banana Leaf Couplet

When the couplet is made into the shape of a banana leaf, it is called a banana leaf couplet. This type of beautifully and uniquely shaped couplet is commonly seen in the private gardens in the south, hung on the corner of a loft, kiosk, gazebo or pavilion.

- **古琴联**

被做成古琴形状的木对联叫“古琴联”，它体现着古代中国人对建筑的雅韵的追求。

Guqin Couplet

The wooden couplet made into the shape of a *Guqin* (ancient Chinese zither) is called a *Guqin* couplet, which represents the pursuit of music-related elegance and charm of architectures of ancient Chinese.

此君联

即竹节对联。其做法是将粗大的竹筒一剖两半，经打磨后在凸面上刻写对联，悬挂于园林、亭榭、书斋等处的室内或室外。这种竹制对联做工很精细，装饰性很强，备受文人、雅士喜爱。因竹子素有“君子”之誉，故称竹制对联为“此君联”。

Jun Couplet

It is namely a couplet on bamboo sections. A thick bamboo tube is split in half, polished, inscribed with couplet texts on its convex side and hung either indoors or outdoors in a garden, pavilion or study. This type of bamboo couplet is sophisticatedly crafted, highly decorative and widely favored by literati because bamboo is generally used figuratively as a gentleman (*Junzi* in Chinese). For this reason, the bamboo couplet is also called *Jun* couplet.

以龙纹为饰的雕花石檐柱叫“龙柱”，一般用于外檐下，为外檐柱。其优点是不怕雨水，又非常精美，起到美化建筑的装饰性作用。石柱的形状有方形、圆形、槽形几种。清代寺观的山门、正殿的檐柱绝大多数是石雕龙柱，龙首有向上、向下之分。若一柱雕有两条龙，便叫“雌雄蟠龙”；若龙首

The eave column made of stone and decorated with loong-patterned carvings is called a "loong column", which is generally placed on the outside of the eave as an outer eave column. It has the advantage of being rain-resistant. Being exquisitely beautiful, it serves the decorative function of beautifying a building. The stone column is shaped in square, circular or grooved form. Vast majority of the eave columns used for the front gates and grand halls of temples in the Qing Dynasty (1616-1911) were stone loong columns. The loong head may be either faced upward or downward. If one single pillar has the carvings of two loongs, it is then called a column of "male and female curled-up loongs". Furthermore, if one of the loongs is facing upward while the other is facing downward, it is then called "heaven turning and earth shattering".

• 曲阜孔庙大成殿雕花石檐柱

Carved Stone Eave Column of the Dacheng Hall in the Confucius Temple, Qufu

一向上，一向下，便叫“翻天覆地”。龙柱上常配以云纹、水纹，以及虾、蟹等水族动物。

以花和鸟为饰的石檐柱叫“花鸟柱”，多见于闽南地区，多为方形石柱，常见的有百鸟朝凤柱和百鸟朝梅柱两种。百鸟朝凤柱纹样为石柱上、下两端各有一只凤凰，并配以繁茂的牡丹花和叶，以及各式各样的鸟。百鸟朝梅柱纹样为一株粗壮的蜡梅盘绕于柱上，千姿百态

Loong columns are frequently coupled with cloud patterns, water patterns and patterns of aquatic creatures such as crabs or prawns.

Stone eave columns decorated with flowers and birds are called "flower and bird column". They are frequently seen in the southern region of Fujian Province. They are usually square pillars. The two most common flower and bird columns are that of "100 birds paying tributes to the phoenix" and that of "100 birds paying tributes to the plum". The pattern of the "100 birds paying tributes to the phoenix" includes two phoenixes on the top and at the bottom of the pillar,

山西王家大院一路连科木雕

该垂花门柱头上的木雕被充分利用了垂花柱有限的空间，雕刻出一路连科图案，造型别致，堪称经典。喻示求取功名的考生可以顺利通过考试，金榜题名。

Wood Carving of "Heron and Lotus Path" with Implied Wish for Success in the Government's Exams All along, at the Wang's Courtyard in Shanxi Province

The wood carving on the pillar tops of the festooned door shows the full use of the limited space on the flower-hanging post to carve the pattern of heron and lotus path in such a delicate style. It is truly a classic masterpiece. The pattern implies the wish that a candidate striving for fame will pass the imperial examinations.

的鸟栖息于树枝上，十分精美。

垂花门麻叶梁头之下有一对倒悬的短柱，柱头向下，酷似一对含苞待放的花蕾，这种短柱被称为“垂花柱”。垂花柱的雕饰十分精美，是垂花门装饰的重要组成部分。

柱础是中国传统建筑构件的一种，俗称“柱顶石”。被置于柱子和台基之间，既可以防潮，又可分担柱子承受的压力。柱础的形式有两大类：一类是单层式柱础，有

• 平遥建筑中的雕狮柱础
Lion Column Base of a Building Located in Pingyao

coupled with lush peony, peony leaves and a multitude of various birds. The pattern of the "100 birds paying tributes to the plum" comprises a thick branch of blooming plum coiling around the pillar, with a variety of birds perching on tree twigs. It is truly delicate and beautiful.

Under the Ma-leave beam head of a festooned door, there is a pair of short upside-down pillars, making the head look like a pair of budding flowers. This pair of short pillars is called "hanging flower columns". The delicately sculpted hanging flower columns is an essential component to the decoration of a festooned door.

The column base is a building component of ancient Chinese architecture, more commonly known as "column top stone". Placed between the pillar and the stylobate, it is moisture-proof and can share the pressure bore by the pillar. There are two types of column bases. One is the single-layered column base, which may take the form of a drum, a basin, floor-paving lotus flowers and animals. The other is the multi-layered one made of two or more overlapping single-layered column bases in different forms.

The carvings on ancient column

鼓式、覆盆式、铺地莲花式、兽式等；另一类是多层式柱础，由两种以上不同形式的单层式柱础重叠而成。

古代柱础上的雕刻有铺地莲花式、仰覆莲花式、覆盆莲花式、海石榴花式、游龙戏水式、宝莲花式、牡丹花式等。其内容更为广泛，包括花鸟、山水、图腾、历史故事、神话等。

base may take the forms of floor-paving lotus flowers, the upward-facing and downward-facing lotus flowers, the downward-facing lotus flowers, the ocean pomegranate flowers, the water-plunging playful loongs, the treasure lotus flowers and the peony flowers. The contents of the carvings are even more extensive, including flowers and birds, landscapes, totems, historical tales and myths.

• 北京圆明园遗址公园中的柱础，雕花十分精美

Beautifully-carved Column Base at the Relics of the Park in the Old Summer Palace, Beijing

• 梯形柱础
Trapezoid Column Base

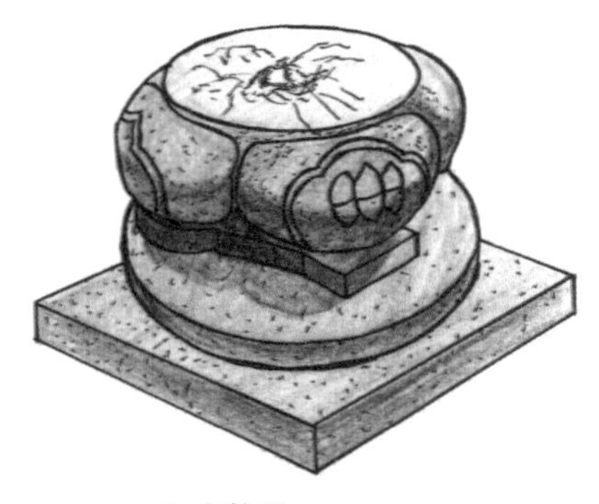

• 如意柱础
Ruyi Column Base

• 卷草重层柱础
Multi-layered Coiled Grass Column Base

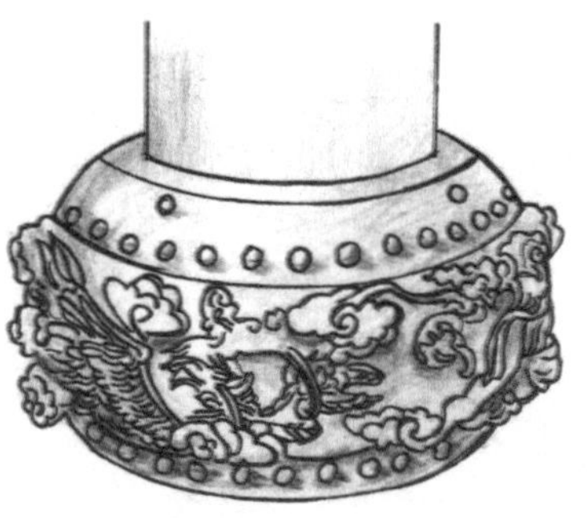

• 云凤柱础
Cloud-and-phoenix Column Base

雀替的制作材料取决于该建筑所用的主要建材，如在木建筑上用木雀替，在石建筑上用石雀替。雀替在北魏时期已具雏形；明清以来，雀替的雕刻装饰效果日渐突出，有龙、凤、仙鹤、花鸟、花

The materials used in the production of the sparrow brace are decided by the primary construction materials used to build the architecture. For instance, a wooden sparrow brace would be used for a wood architecture and a stone sparrow brace would be used for a stone

篮、金蟾等形式，雕法则有圆雕、浮雕、透雕。

牛腿的基本形状犹如上大下小的直角三角形，但也有少量是方形的。它依附在檐柱外侧的上端，其上方直接或间接地承载着屋檐的重量。牛腿上的图案装饰十分丰富，且寓意深刻，包括动物题材、人类题材、山水题材，等等。例如，牛腿上常雕刻有狮子图案，古人认

architecture. The sparrow brace was in its infancy during the Northern Wei Dynasty (386-534). Its decorative effect became increasingly prominent since the Ming and Qing dynasties (1368-1911). Some of the forms used for a sparrow brace include the loong, phoenix, crane, flower and bird, flower basket and toad. Some of the sculpture methods include circular carving, relief carving and hollowed-out carving.

• 精致的彩色雕花雀替
Exquisitely Carved and Colored Sparrow Brace

清代东阳木雕九狮纹牛腿（图片提供：FOTOE）
Dongyang Wood Carving of the Nine-lion Pattern on the Corbel in the Qing Dynasty (1616-1911)

为狮子是瑞兽，可以保平安。福、禄、寿三星，渔樵，耕读等也常出现在牛腿上，此外还有山水画，梅、兰、竹、菊，松柏，石榴，莲荷，牡丹，等等。

The basic shape of the corbel is a right-angled triangle with a wide top and narrow bottom. There are also a few of them in the shape of a square. It is attached to the upper and outer end of the eave column to directly or indirectly carry the weight of the eave above it. The corbel is rich in decorative patterns with meaningful themes such as animals, people or landscapes. For example, sometimes the corbel is carved with a lion on its surface. Since people in ancient times believed that lions were auspicious animals, these carvings were intended to express hope for peace. Patterns of the "three stars of happiness, fortune and longevity" and "lives of fishery, logging, farming and studying" also frequently appear on corbels. There are also patterns of landscape painting with meandering mountain ranges, as well as "plum blossom, orchid, bamboo and chrysanthemum" "pine trees, furs and pomegranate" and "lotus and peony flowers".

> 梁枋

梁是屋顶木构架中横向负重的构件之一，也被称为“柁”。中国传统建筑以土木结构为主，梁是建筑必不可少的构件。根据部位的不同，梁有很多专用名称。例如，在攒尖式屋顶的木架构中，不直接交会在立柱上，用于加高和收缩屋顶

> The Roof Beam

The beam, also known as the "girder", is one of the horizontal weight-bearing components in the wood framework of the roof. Traditional Chinese architecture was dominated by wood and earth structures and the beam was an indispensable structure. The beam has various specific names in accordance

• 北京颐和园里的抹角梁
Blunt-corner Beam at the Summer Palace, Beijing

的梁叫“抹角梁”。梁不仅仅具有房屋的支架的实用功能，也具有极强的美化装饰功能。传统建筑中的梁常被绘上各种精美的图案，既可以防止梁受潮变形，又成为一种精美的点缀。

枋是被安装在两柱之间或两柱上端的扁方形木材，两个柱子的上端被榫卯连在一起。有斗拱的建筑，要专设几根枋，用来承载、拉拽、压住斗拱。枋有时也被安装在檩（又叫“桁”“栋”，被架设在枋木之上或前后斗拱之间）和梁的

穿插枋

穿插枋用于抱头梁或挑尖梁的下面。作用是承载抱头梁或挑尖梁，增强抱头梁或挑尖梁的承重能力。

Interjecting *Fang*

The interjecting *Fang* is used underneath the head-embracing beam or the tapering beam to support them and strengthen their load-bearing capacity.

with where it is positioned. For example, in the wood structure of a pyramidal roof, the beam is not directly placed on the top of pillars. Those used to raise the height of the roof and taper off the top of the roof are called "blunt-corner beam". Beams have not only the practical function of supporting houses, but also highly decorative functions. In traditional architectures, beams are often painted with a variety of beautiful patterns in order to prevent deformation caused by dampness and to provide exquisite embellishment.

The *Fang* is a flat square timber installed horizontally between two pillars or on the top of two pillars. Through the use of tenons, the *Fang* is connected with the top of two pillars. Those buildings with wood brackets require specifically designed *Fang* to support, pull or press the wood bracket. *Fang* can also be installed below the ridgepole (also known as the "purlin", is mounted above the *Fang* or in between the front and back bracket set) and beam to strengthen the load-bearing capacity of ridgepoles and beams and prevent their deformation. *Fang* is given different names according to the specific locations they are installed.

下面，有增强檩和梁的承重能力，防止檩和梁变形的作用。枋木因安装位置不同而有不同的名称。中国传统建筑以木构架为主，枋是屋顶的重要支撑部分，同时也是彩画装饰的重点部位。

Traditional Chinese architectures are primarily built of wooden structures, and the *Fang* is not only an important supporting part of the roof, but also a key part for color-painted decorations.

• **脊枋**

脊枋即位于房屋脊柱之下的方形木材，位于屋架的最高处，其上便是脊檩。脊枋的作用是承载脊檩，增强脊檩的承重能力。

Ridge *Fang*

The square timber under the spine pillar of a building, and on the highest part of the roof frame is called the ridge *Fang*. The ridge *Fang* is used to support and strengthen the load-bearing capacity of the ridge purlin.

• **抱头梁和挑尖梁**

位于檐头部位的梁，一种是抱头梁，另一种是挑尖梁。

Head-embracing Beam and Tapering Beam

The tapering beam and the head-embracing beam are the exposed portion of the beam at the end of the eaves.

• 大、小额枋

额枋指檐柱之间的方形木材。 有的建筑有两根额枋， 通常上面的一根稍大，叫“大额枋”，下面的一根较小，叫“小额枋”。

Big and Small Architraves

The architrave refers to the square piece of wood between eave columns. Some buildings make use of two architraves. The slightly larger one on top is called the "large architrave" while the slightly smaller piece below is called the "small architrave".

• 坐斗枋

坐斗枋是位于大额枋之上的一条方木，用于承载斗拱，斗拱最底下的大斗就“坐”于平板枋上，故叫“坐斗枋”。

Bracket-seat *Fang*

The bracket-seat *Fang* is a piece of square wood located above the large architrave to support the wood bracket, and the big bracket at the bottom sits directly on this flat panel of *Fang*, called the "bracket-seat *Fang*".

> 护栏

栏杆有木栏杆、石栏杆、石木栏杆、矮墙等种类，被安装在高台、台基的边缘，或被装于两柱之间或窗下，以代替半墙，多见于临水建筑、楼阁、走廊等处。

木栏杆多用于临水建筑、楼阁、走廊等处。竖木为栏，横木为

> The Guard Rail

The railing comes in various forms, such as the wood railing, the stone railing, the wood and stone railing, low walls, and so on. It is installed on the border of tall decks and stylobates, in between pillars, or beneath windows as a replacement for low walls. It is commonly seen on those buildings near waterfronts, as well as

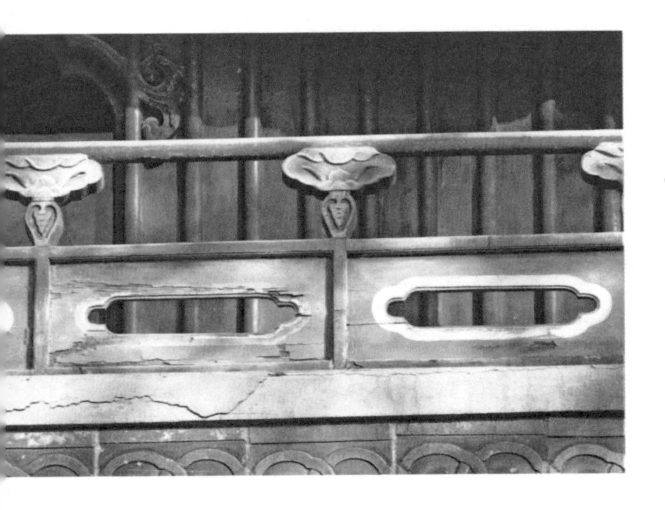

• 寻杖木栏杆

寻杖木栏杆是最常见的一种栏杆形式，多用于檐下栏杆和楼梯两侧扶手。其主要部件有望柱、寻杖、扶手、腰枋、下枋、地袱、绦环板、牙子、荷叶净瓶等。

Wood Staff Railing

The wood staff railing is the most commonly seen railing used as handrails underneath the eaves or on both sides of a staircase. Its main components include the baluster, the handrails, the waist *Fang*, the lower *Fang*, the ground rail, the relief-line board, the teeth, the lotus-leaf vase, etc.

杆，为防护而设。木栏杆有高、低两种，较高的多为寻杖栏杆、靠背栏杆（也称“美人靠”“吴王靠”），此外还有花式栏杆和瓶式栏杆。

- **花式栏杆**

  花式栏杆表现为寻杖栏杆的变形和简化。人们通常不对花式栏杆采用寻杖形式，而是在整个栏杆板上雕刻各种纹样，有冰裂纹、盘长纹、龟背锦等。用料较轻巧，花式变化丰富，样式别致。

  Flower-patterned Railings

  The flower-patterned railings are a simplified variation of the wood staff railing. The flower-patterned railing generally does not be applied a staff, but the railing boards are carved with a variety of patterns instead, such as ice-crack patterns, Buddhist lucky patterns and turtle-back brocade patterns. The materials utilized for flower-patterned railings are light with various patterns and unique styles.

pavilions and corridors.

Wood railing is generally used on buildings near waterfronts, as well as pavilions and corridors. The vertical wood pieces are called *Lan*, and the horizontal pieces are called *Gan*. It is a protection design installed between two pillars. There are two kinds of wood railings: the tall one and the low one. Most of the taller railings are staff railings (*Xunzhang* railings), backrest railings (*Kaobei* railings), also known as the "beauty backrest" or "King Wu's backrest", the flower-patterned railings and the vase-shaped railings.

靠背栏杆

靠背栏杆又名“吴王靠”“美人靠”“飞来椅”，是一种形如椅子靠背的栏杆，高度在50厘米左右，长度依所属建筑开间的尺寸而定。常用于园林的亭、榭、轩、阁等小型建筑的外围，既可用作栏杆，亦可供人休息，安全而舒适。

Backrest Railings

The backrest railing, also known as the "King Wu's backrest" "beauty backrest" or "the chair flew from afar", is a type of railing with a chair's back for backrest. It is approximately 50 centimeters in height, with the length set relative to the span of the construction. It is frequently seen in the outer rings of lofts, kiosks, gazebos, pavilions and other small structures in the garden. It not only serves as a railing but also provides people with a place to rest safely and comfortably.

瓶式木栏杆

瓶式栏杆是清代受外国建筑的影响而出现的一种西洋式栏杆。栏板采用由多根木料旋成的瓶式直棂条，而不是栏板形式。瓶式栏杆经过匠人的彩绘，装饰性更强。

Vase-shaped Wood Railing

The vase-shaped railing is a type of Western-styled railing influenced by foreign architecture in the Qing Dynasty (1616-1911). The rail board is made of multiple vase-shaped vertical bars by cutting the wood materials, instead of the traditional rail board. Color-painted by craftsmen, the vase-shaped railing becomes more decorative in its attributes.

石栏杆又叫“勾栏”，用于殿堂、亭榭、月台的须弥座式高台或台基上，也用于石桥、华表、碑、园中水岸边，其用途是维护安全。石栏杆具有极好的装饰性，还有分隔或连接不同景区的作用，能够丰富园林的空间层次感。

Stone railings, also known as "*Goulan*" in Chinese, are used as safety precaution on the high-raised platform or stylobate with the Buddhist Sumeru base as seen in palace halls, pavilions and moon platforms as well as on stone bridges, ornamental columns, monuments and banks of water in gardens. However, since stone railings have excellent decorative effects and can separate or connect different scenic spots, they are often used to enrich the layered space in the courtyard.

• **北京恭王府花园石栏杆**

石栏杆由很多小单元组成，每个小单元包括地袱、栏板、望柱，起头或收尾处还被加上了抱鼓石。

Stone Railing in the Flower Garden of Prince Gong's Mansion, Beijing

The stone railings are composed of numerous small units. Every unit includes such elements as ground rail, rail board and baluster, with drum-holding stones at the beginning and the end of the railings.

望柱

望柱也称“栏杆柱”，是位于栏杆的栏板和栏板之间的短柱。望柱分柱头和柱身两部分，柱头部分有很多雕刻性装饰，常见于宫殿、园林、桥梁等建筑中。

Baluster

The baluster, also known as the "railing baluster", is a short pillar located between railing boards. A baluster is composed of a baluster head and a baluster body. The baluster head is usually carved with numerous decorations, as seen so commonly in the architecture of palaces, courtyards, bridges, and so forth.

• 石望柱
Stone Baluster

• 铜望柱
Copper Baluster

石望柱的柱头因装饰纹样不同而有很多种，使用必须符合规定。其官式做法有蕉叶柱头、云龙柱头、凤柱头、狮子柱头、八不蹭柱头、莲瓣柱头、覆莲柱头、石榴柱头、二十四节气柱头、风云柱头、仙人柱头、叠落云子柱头、马尾柱头等多种形式；民间的石望柱柱头的纹样则比较随意。

Since the baluster head of a stone baluster can be decorated with various patterns, its usage must comply with the regulations. The typical baluster heads designated for official usage include banana leaf baluster head, cloud loong baluster head, phoenix baluster head, lion baluster head, eight no-rubbings baluster head, lotus-petal baluster head, inverted lotus baluster head, pomegranate baluster

• 北京北海公园的石榴望柱头

Pomegranate Baluster Head in Beihai Park, Beijing

• 北京北海公园琼岛莲花望柱头

Lotus Baluster Head on the Jade Islet in Beihai Park, Beijing

head, twenty-four solar terms baluster head, wind and cloud baluster head, the immortal baluster head, cascading cloud baluster head, horse tail baluster head, and so forth. In contrast, the stone baluster in the civilian context has more free usage of the patterns for baluster head.

- **北京北海公园琼岛延楼长廊云龙纹望柱头**

云龙纹望柱因柱头上装饰有云龙纹而得名。工匠们充分利用柱头有限的空间，用祥云衬托龙的威势，龙身周围饰以云纹，对龙纹采用浮雕形式，龙身上的鳞纹清晰可见。

Cloud Loong Baluster Head of the Corridor of Yanlou Tower on the Jade Islet in Beihai Park, Beijing

The cloud loong baluster head got its name from its patterns of loong and cloud on the baluster head decoration. The craftsmen fully utilized the limited space on the baluster head to depict the loong and support its powerful stance by the usage of patterned cloud. The body of the loong is also adorned with clouds. The loong is embossed on the surface with its scale patterns clearly visible.

> 台基

台基是建筑的基座，包含房屋的基础、柱顶石等。中国传统建筑很重视台基，把台基、墙体、屋顶视为建筑的三大组成部分。台基分

• 北京故宫太和门石狮座下的须弥座
Buddhist Sumeru Base under the Stone Lions of the Gate of the Supreme Harmony in the Forbidden City, Beijing

> The Stylobate

The stylobate serves as the foundation base of a building, and contains the groundwork, column base and others. Traditional Chinese architecture attaches great importance to the stylobate, and considered the stylobate, the wall and the roof as the three major components of a building. Stylobates are divided into the basic stylobate and the Sumeru stylobate. Moon platform and balustrades are also important components of a stylobate.

Stylobates may differ in accordance with the rank of their respective architecture. In ancient times, officials had strict regulations and rules as to the use of stylobates. According to the *Records of the Qing Dynasty* (*Daqing Huidian*), houses that belong to the officials of the level below a prince but above Level Three in rank should

- **龙头须弥座**

此须弥座带有石栏，其望柱之下、须弥座的上枋部位被安装了一排石雕龙头，位于角部的龙头较大，为爬出状，龙嘴处有一个排水孔，可排泄须弥座台面上的雨水。

Loong Head Buddhist Sumeru Base

A row of stone loong heads is set on the upper *Fang* under the baluster, surrounded by stone railings. The loong heads located at the corners are larger in size and posed in a crawling form. A drain hole is located at the back of the loong's mouth, which serves to drain the rainwater from the Buddhist Sumeru base.

为普通台基和须弥座式台基，月台和栏杆也是台基的重要组成部分。

台基因建筑物的级别而有所不同。古代官方对台基的使用有严格的规定，据《大清会典》记载：公侯以下、三品官以上的房屋台基准高二尺，四品官以下到市民的房屋台基高一尺。

须弥座为中国传统建筑中的台基所广泛运用，用以衬托建筑物的神圣和崇高。须弥座的体积虽有大小，但基本形态是相同的。须弥座从上到下的各部分分别为上枋、上枭、束腰、下枭、下枋、圭角等。

have the standard stylobate at two *Chi* (two-thirds of a meter) high. Houses that belong to Level-Four officials and downward to civilians should have the standard stylobate at one *Chi* (one-third of a meter) high.

Known as the Buddhist Sumeru base, the base is used to enhance the holiness and loftiness of an architecture, and thus has been widely used as the stylobate in traditional Chinese architecture. Although they may vary in size, Buddhist Sumeru bases are identical in shape. The various parts of the Buddhist Sumeru base are categorized from top to bottom into upper *Fang*, upper cyma, waist, lower cyma, lower *Fang*, etc.

束腰处有椀花结带纹的须弥座

因汉白玉须弥座用于不同等级的建筑，故汉白玉须弥座被施以的雕花处理也不同，最简单的一种是仅在束腰部位进行椀花雕刻，其余各处均无雕花。椀花结带纹是一种上下对称的纹样，动感很强。

Pattern of Splashing Tea Foam with Ribbon Band on the Waist of the Buddhist Sumeru Base

Since the white marble Buddhist Sumeru base are used in architecture of different classes, the carving patterns used for white marble Buddhist Sumeru base are also differently handled. The simplest pattern is the use of splashing tea foam pattern carvings on the waist portion of the Buddhist Sumeru base, while the remaining portions are kept in plain surface. The pattern of splashing tea foam with ribbon band is vertically symmetrical which shows a very strong dynamic sense.

须弥座四角位置的喷水兽

喷水兽的实际作用是排水，它是一种龙首兽身的瑞兽，因大部分只被做出头部，又叫“螭首”。位于须弥座四角位置的瑞兽形体较大，露出身体的一半，形象也很生动，叫“大龙头”或“四角龙头”。其他部位的龙头都较小，叫“小龙头”或“正身龙头”。

Water-sprinkling Beasts at the Four Corners of the Buddhist Sumeru Base

The water-sprinkling beast is practically used as a water outlet. It is an auspicious animal in the shape of the head of a loong with the body of a beast. Since usually only its head is exposed outside, it is also called "*Chishou* (the head of a hornless loong)". The water-sprinkling beasts at the four corners of the Buddhist Sumeru base are larger in size, with only half of their bodies exposed in vivid forms. They are called the large loong head or four-corner loong head. The loong heads on the other parts of the Buddhist Sumeru base are smaller in size and called the small loong head or full-front loong head.

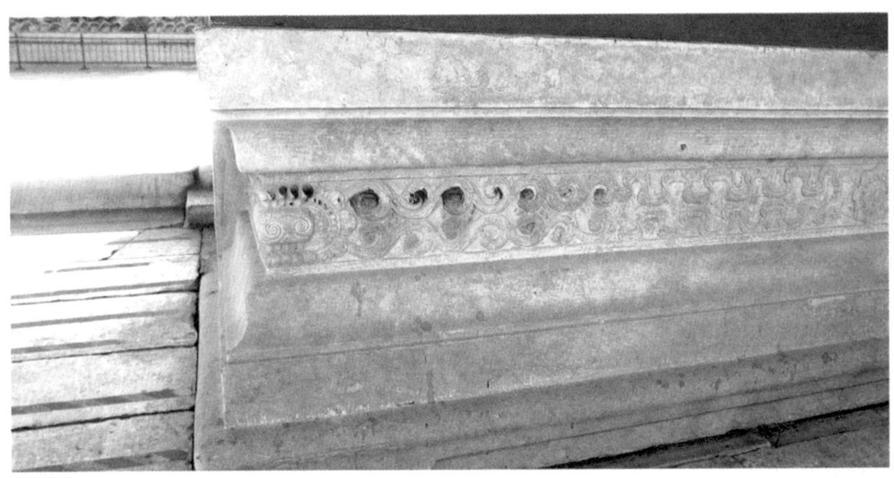

- **束腰和上枋处有雕花的须弥座**

这种汉白玉须弥座的雕饰较多，一般出现在束腰和上枋上。束腰部位的纹饰仍是椀花结带纹，上枋上雕刻有卷草纹。

Buddhist Sumeru Base with Carvings on Its Waist and Upper *Fang* Section

This type of white marble Buddhist Sumeru base generally has more carvings on its waist and upper *Fang* sections. The carving on the waist is the pattern of splashing tea foam with ribbon band while the carving on the upper *Fang* section is the curled grass pattern.

- **全雕花的须弥座**

这种汉白玉须弥座的所有部位上都有雕花。束腰部位的纹饰仍是椀花结带纹，上、下枋多雕刻有卷草纹、宝相花纹或云龙纹，上、下枭多雕刻有莲花瓣纹。

Fully-carved Buddhist Sumeru Base

This type of Buddhist Sumeru base is fully carved throughout its bodies. The pattern of splashing tea foam with ribbon is still used on its waist, while the curled grass, composite flowers and cloud and loong patterns are used on its upper *Fang* and lower *Fang* sections. Its upper cyma and lower cyma are often carved with lotus-petal patterns.

> 铺地

传统建筑装饰中对地面的装饰手法主要为铺地。铺地就是通过镶嵌、拼接的方法构成各种各样的花纹、图案，以增强地面的视觉效果。

铺地又叫“墁地”，指用砖石

• 散水
Apron

> The Ground Paving

The floors in traditional architecture are primarily decorated through the use of ground paving techniques, where various types of patterns and pictures are inlaid or pieced up together to increase the visual effect of the ground.

Also known as "*Mandi*", ground paving refers to the brick or stone pavement on the indoor or outdoor ground, including the roads in the courtyards and gardens. Indoor ground paving refers to the pavement on the indoor ground where square bricks or long bricks are generally used. There are three kinds of courtyard ground paving: apron, paved path in courtyards or gardens and paved square. Among the three, paved path in courtyards or gardens involves more varieties of skills. Apron is the brick pavement made on the ground

长砖墁地砖缝

用砖墁地，会因砖的排列方式不同而有不同的缝隙，即“砖缝”。砖缝其实是一类图案，常见的有十字缝、拐子锦、褥子面、人字缝、丹墀、套八方等。

Brick Gaps of Long-brick Paved Square

In bricked paved square, different gaps may form as result of the various ways of arranging the bricks. These gaps are called brick gaps. In fact, brick gaps are also a type of pattern. Some of the frequently seen brick gaps include the cross gaps, brocade of cane gaps, quilt surface gaps, "人" character gaps, red palace-steps gaps and eight-direction joint gaps.

海墁

Paved Square

铺设室内外的地面，也包括庭院和园林中的道路。室内墁地指建筑内部地面的铺装，一般用方砖或长砖墁地。庭院墁地有散水、甬路、海墁三类。其中，甬路的做法较多。散水是在屋檐下、台阶旁沿着前、后檐及山墙在地面上墁砖。墁砖时要略留有坡度，让屋檐下的雨水尽快流走，房屋四周没有积水，这样

along the front eave, back eave, and gables beneath the eaves or beside the stairs. Brick pavement is slightly sloped to facilitate the flow of rainwater so that no water will be accumulated around the building. It not only protects the groundwork from water erosion, but also beautifies the ground around the building. The frequently used apron includes the patterns of first-grade volume and serial

• **雕花甬路**

雕花甬路主要指在甬路两侧铺散水的位置改用有雕花的方砖，或用瓦片组成好看的纹饰，也可采用多色卵石摆成图案。

Paved Path in Courtyards or Gardens with Carved Patterns

Paved path in courtyards or gardens with carved patterns refers to the apron area on both sides of the paved path in courtyards or gardens being replaced by pattern-carved square bricks or attractive patterns of pieced-up tiles. Sometimes, multi-colored pebbles are also used to create patterns.

便可以保护地基不受雨水的侵蚀，也有美化建筑周围地面的作用。常用的散水铺法有一品书、连环锦等。另外，甬路、御路的两侧也要有散水。甬路是庭院中的主要交通路线，一般以方砖或长砖墁地。甬路分大式做法和小式做法。宫廷中的甬路称为“御路”，以条石墁地，两侧也要以砖铺设散水。海墁指在散水、甬路之外，在庭院中

brocade. Sometimes, apron is also used on both sides of paved path in courtyards or gardens and the imperial paths. Paved path in courtyards or gardens is the main traffic route in a courtyard. It is usually paved out of square bricks or long bricks. Paved path in courtyards or gardens can be built in big-type or small-type approaches. The paved path in courtyards or gardens in palaces, paved with bar stones, is called the imperial path. It also

其他地方也都用条砖墁地的做法。“海墁”是老北京话，“海”是全部、没有边界之意。业内有“竖墁甬路横墁地”的口诀，是说海墁的条砖要沿东西方位放置。

花街铺地是园林铺地的种类之一，多见于江南园林。花街铺地就是用砖、瓦、石片、印石、瓷缸片等材料在地面上铺成各种图案、花纹。它的种类很多，大多是就地

方砖卵石嵌花路

这是北方宫苑中常用铺地方法。一般在道路中间铺方砖，在方砖两侧铺卵石带，以砖、瓦分段，将卵石拼成“寿”字、如意、铜钱、扇形、海棠等图案，再以各色卵石填心。

Patterned Road Inlaid with Square Bricks and Pebbles

This is a very common ground paving practice used in palaces and courtyards in the north. Square bricks are generally paved in the middle of the road, with pebbles paved along both sides. Brick and tile fragments are used to outline the pebbles in the forms of the Chinese character "寿 (longevity)", *Ruyi*, coin, fan, begonia or others. Then pebbles of different colors are used to fill in the centers.

has apron installed on both sides. Paved square refers to the bar-brick-paved space in courtyards other than the apron and paved path areas. The paved square (*Haiman* in Chinese) is an old Beijing term: "*Hai*" means all, or being without borders. In the industry, the mnemonic jargon "Paved path in courtyards or gardens in vertical directions and paved square in horizontal directions" refers to the fact that the bar bricks used for paved square should be laid in east and west orientation.

Flower-street ground paving is a type of garden ground paving frequently seen in gardens in southern China. It's the combined use of bricks, tiles, stone plates, pebble stones and ceramic fragments to create various types of pictures and patterns for the ground pavement. There are many kinds of flower-street ground paving, but most of them make use of local materials. The low-value materials may achieve high-class effects and accomplish high degrees of craftsmanship and artistry. For instance, a simple use of bricks and tiles may compose basket pattern, whorl pattern, chessboard pattern and " 人 " character pattern. The use of bricks can also form such frames as hexagons, hexagonal frames, hexagonal

取材，低材高用，体现了较高的技艺水平。如纯以砖瓦摆成席纹、斗纹、间方纹、“人”字纹等；以砖瓦做成六角、套六角、套六方、套八方等方框，在框内填充以各色卵石或碎瓷片；亦可以卵石与瓦混砌，做成套钱、芝花等纹；还有以各色卵石铺地的做法，花纹如织锦，色彩对比鲜明。

direction, and octagonal direction and fill the inside of the frame with a variety of colored pebbles and porcelain pieces. Furthermore, pebbles and tiles can also be mixed in use to form coin frame and flower patterns. Such ground paving as paved with colored pebbles may display patterns like brocade in a sharp contrast of colors.

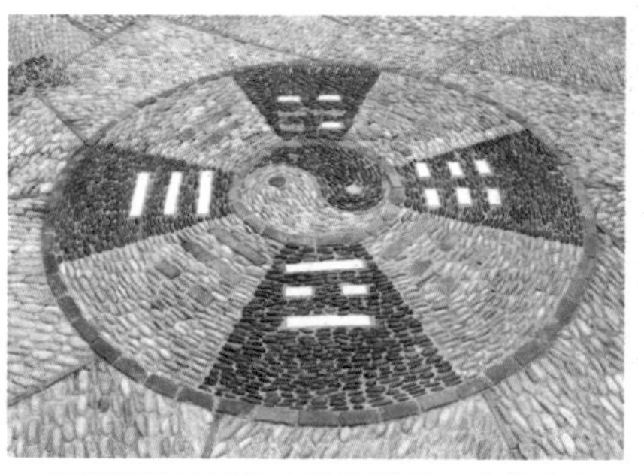

• 苏州狮子林里的八卦纹铺地，有给人们带来吉祥之意

Eight Diagrams-patterned Ground Paving in Suzhou Lion Grave Garden, Believed to Bring in Good Luck

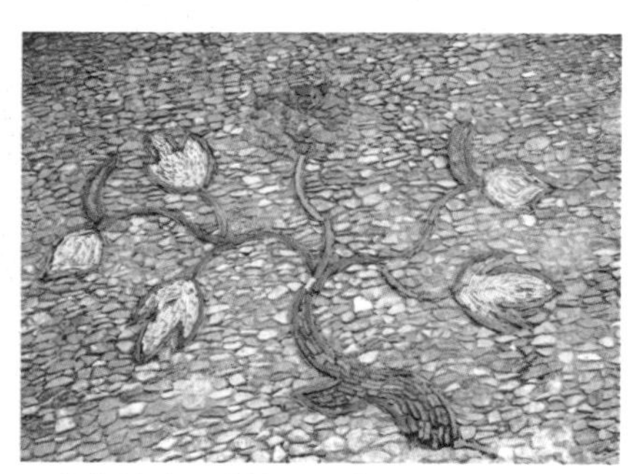

• 花街铺地之植物纹图案

Vegetation Pattern in Flower-street Ground Paving

• 吉祥如意铺地

双鱼的鱼鳞、尾巴被突出雕刻，形象生动。

Ground Paving Meaning Good Luck and Happiness

The scales and tails of two fish are vividly carved.

室内铺地

建筑内部地面的铺装一般采用方砖或长砖，从工艺上分细墁和粗墁两种。还有一种最高级的墁地方式叫“金砖墁地”，仅用于最高级别的宫殿中。

细墁是一种高级的砖铺地面工艺，所用方砖经过砍打、细磨，仔细墁平后，要被施以表处理，包括清理砖体表面、用磨头蘸水再次将砖体打磨平整，待干后再用生桐油反复擦拭。为防潮及减少起沙，一般都要铺设水磨方砖。

• 细墁铺地
Fine Ground Paving

粗墁是一种砖铺地面工艺，所用砖未经过砍打、细磨处理，但墁砖时工艺与细墁相同，只是最后不抹油灰，也不用生桐油，只用白沙子灰勾缝。

金砖墁地，金砖是一种用澄浆泥烧制而成的大方砖，规格为50 厘米 × 50 厘米 × 8 厘米，是专门为宫廷烧制的方砖。这种砖质地坚硬，表面有光泽，敲之有金属声，故称“金砖”。

The Indoor Ground Paving

The indoor ground surface is generally paved with square bricks or long bricks. The indoor ground paving can be classified into the rough ground paving and fine ground paving by their respective craftsmanship. There is also a ground paving of the highest class, called the "gold brick ground paving", which is seen only in the hall of the highest-class imperial palaces.

Fine ground paving is a high-class type of brick pavement. The square bricks utilized in this ground paving are cut and sanded, then carefully leveled after being paved, and processed on the surface by clearing and water milling the surface. They will then be polished and cleaned before

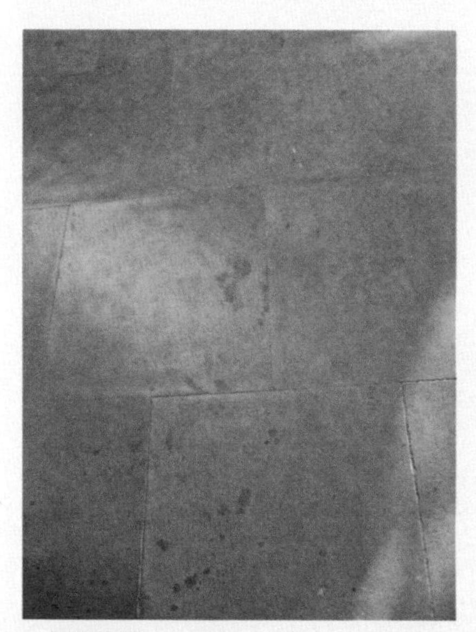

• 粗墁铺地

Rough Ground Paving

they are repeatedly rubbed with raw tung oil after they are dried. The square bricks are often water milled in order to make them moisture-proof and less friable.

Rough ground paving is a type of brick ground paving that makes use of uncut and unpolished bricks. Although the paving process is the same as the fine ground paving, the rough ground paving process does not wipe away excess putty, nor use raw tung oil at the end, yet uses fine white sand to fill the joints between bricks.

The gold brick ground paving uses a type of large square brick made and fired from clear clay pulp. With a dimension of 50 cm × 50 cm × 8 cm, these square bricks are specifically produced for palaces. Since they are solid, hard, polished and sound like metal when struck, they are called "gold bricks".

• 北京故宫乾清宫内的金砖墁地

Gold Brick Paved Ground at the Palace of Heavenly Purity in the Forbidden City, Beijing

> 室外建筑的其他装饰

石碑

宫殿、寺观前常设有石碑，石碑由碑首、碑身、碑座三部分组成。碑首，早期有圆首形和尖首形两种，这大约与周代琬圭和琰圭两种礼玉有关。碑身是碑的主体部分，一般以竖长方形的巨石雕刻而成。碑的正面叫“碑阳”，一般以刻文字为主，也有的刻有山水、人物、花草、禽鸟、地图等。碑的背面叫“碑阴”，多刻有诸如门生、故吏和出资建碑者的姓名等文字。碑侧也刻有文字，以及精美的图案、花纹。碑座是安放碑身的石座，被称为“趺”。早期的碑座为简单的长方形，四面各有朱雀、玄武、青龙、白虎“四灵神”。南北

> Other Decorations of Outdoor Architectures

The Stone Tablet

The stone tablet is usually set up in front of palaces and temples. A stone tablet is composed of the tablet head, the tablet body and the tablet base. The earlier tablet heads were circular or pointy in shape, which may have something to do with the two types of ritual jade, jade tablet with round head and jade tablet with pointed head, from the Zhou Dynasty (1046 B.C.-221 B.C.). The tablet body is the main part of a stone tablet, which is generally carved out of a vertical rectangular boulder. The front of the tablet, called "*Beiyang*", is typically carved with texts and sometimes with landscapes, personages, flowers, birds, maps or other items. The back of the tablet, also known as "*Beiyin*", is often

成都清羊宫门前石碑

石碑下的龟形兽为赑屃，是龙的九子之一，形状似乌龟，好负重，身上驮有石碑。常见于寺庙或祠堂建筑中。碑文两旁的文龙是负屃，它平生好文，所以多被用来装饰碑文两侧。

Stone Tablet in Front of the Qingyang Palace, Chengdu

The tortoise-like beast beneath the stone tablet is called "*Bixi*" in Chinese, one of the loong's nine sons. It is tortoise-shaped and has a great load-bearing capacity. It has a stone tablet mounted on its back and is frequently seen inside temples and ancestral halls. The loongs on both sides of the inscription are called "*Fuxi*". Since *Fuxi* enjoyed literature, they are often used as decorations by the side of the inscriptions.

朝时期出现了龟形碑座，被称为“龟趺”，将碑座改制成为龟形，有期望石碑被永久保存之意。

铜铸饰品

宫殿、寺庙前常有龟形、鹤形、龙形等的铜铸饰品。摆放铜铸饰品既增强了建筑环境的美观性，又可以显示身份的尊贵。除此之外，还有铜坐狮、铜凤、铜麒麟、铜象、铜牛等，除了铜牛，一般均以成对、成组的方式陈设。

carved with the names of the followers, disciples or individual who funded the construction of the tablet. The sides of the tablet may also have carved texts and exquisite patterns. The tablet base, also known as the "*Fu*", is the stone base on which the tablet body is mounted. Earlier tablet base were made into simple rectangular shapes with the four spiritual creatures, red sparrow, dark turtle, green loong and white tiger, carved on each of its four sides. The tortoise-shaped tablet base, called "*Guifu*", appeared during the Southern and Northern dynasties (420-589), with the tablet bases essentially shaped like a tortoise. They symbolized the permanence of the stone tablet.

• 北京故宫太和殿前的铜鹤

Bronze Crane in Front of the Hall of Supreme Harmony in the Forbidden City, Beijing

The Cast Bronze Decoration

The front of palaces and temples is often decorated with bronze decorations cast in the form of tortoises, cranes, loongs, and so on. Bronze decorations not only strengthen the aesthetic beauty in the environment of a building, but also symbolize its noble status. Beyond that, there are also bronze lions, phoenixes, kylins, elephants and bulls among other creatures. Aside from bronze bulls, most of these statues are generally displayed in pairs or sets.

• 北京故宫乾清宫前的铜龟

Bronze Tortoise in Front of the Palace of Heavenly Purity in the Forbidden City, Beijing

日晷

日晷是古代的时钟，晷盘面平行于赤道面，晷针垂直于晷盘面。它利用太阳的投影和地球自转的原理，借指针所生阴影的位置来显示时刻。古代宫殿、园林等建筑的空地处常放有日晷，方便人们知道时间。

The Sundial

The sundial was used to tell the time in ancient times. The surface of the sundial disk is parallel to the equatorial plane and the dial needle is set perpendicular to it. Using the principle of solar projection and the rotation of the earth, the shadow of the dial needle will display the time by the marked scale. The sundial used to be placed in the open space in front of such buildings as ancient palaces and courtyards to make it convenient for people to determine the time.

• 北京故宫太和殿前的日晷

Sundial in Front of the Hall of Supreme Harmony in the Forbidden City, Beijing

鼎式香炉

The *Ding*-style Incense Burner

鼎式香炉也叫“香炉鼎”，从古代一直沿用至今，可用来燃烧檀香和松枝，也被认为可用来祈求吉祥。周代有“天子九鼎，诸侯七鼎，卿大夫五鼎，元士三鼎”等使用数量的规定。

The *Ding*-style incense burner, also known as the incense burner *Ding*(cauldron), has been used since ancient times. It is used for burning sandalwood or pine log, and is also believed to bring forth good luck. During the Zhou Dynasty), the use of *Ding* was regulated as "Nine pieces of *Ding* are used by the emperor, seven pieces of *Ding* used by dukes and princes, five pieces of *Ding* by patriarchal clans, and three pieces of *Ding* by lower officials".

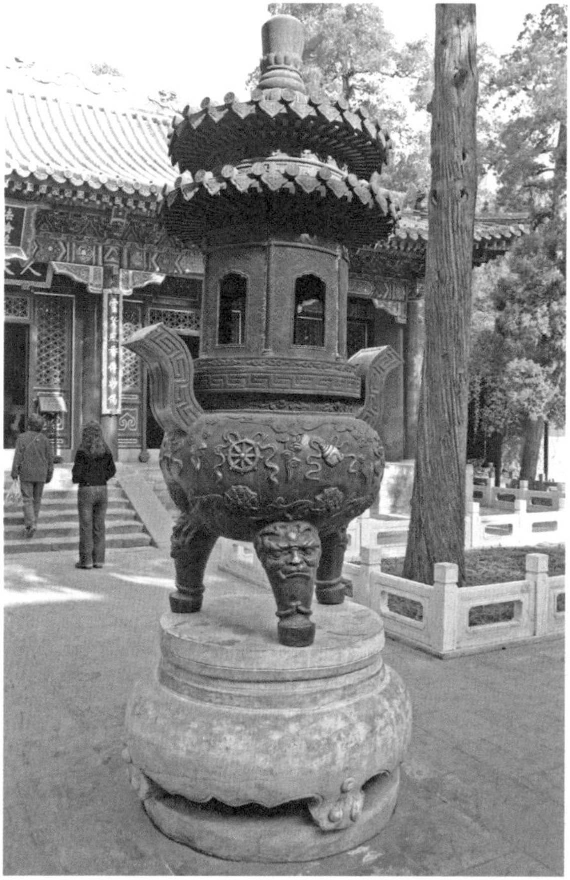

• 北京故宫太和殿鼎式香炉

Ding-style Incense Burner at the Hall of Supreme Harmony in the Forbidden City, Beijing

传统建筑室内装饰

The Interior Decoration of Traditional Architecture

室内装饰是中国传统建筑装饰不可忽略的重要部分，讲究层次分明、有序、协调。传统建筑的室内装饰包括挂落、门帘架、罩、天花、藻井等构件，这些构件的装饰功能几乎与其结构功能同样重要。受技艺、地域、民俗等因素影响，传统建筑的室内装饰风格各有不同。

As an important part of Chinese traditional architecture that cannot be neglected, interior decoration puts much emphasis on structured layers, orderliness and coordination. The interior decoration of traditional architecture is mostly manifested in hanging fascia, door curtain frame, mask, ceiling patterns, caisson and other components. The decorative functions of these components are almost as equally important as their structural functions. Influenced by such factors as craftsmanship, geographic region, folk custom, and so on, the interior decoration of traditional architecture varies in styles.

> 挂落

挂落是额枋下的一种建筑构件，常以镂空的木板或雕花板做成，也可由细小的木条搭接而成，

> The Hanging Fascia

The hanging fascia is a construction component set below the architrave. It is usually either made of hollowed wooden frame or carved panels, or composed of small strips of wood for the purpose of dividing the interior space. More often than not, hanging fascia is the focus of the architectural decoration with openwork carving or colored treatment. The basic patterns of hanging fascia include steps of prospect, lantern frames,

• 山西乔家大院第二院敦品第大门挂落八骏图
Hanging Fascia with the Pattern of Eight Spirited Horses, Hung from the Gate of the Second Courtyard of the Qiao's Courtyand in Shanxi Province

• 双凤朝阳木雕挂落
Wooden Hanging Fascia Carved with Patterns of Two Phoenixes Darting toward the Sun

用来划分室内空间。同时，挂落在建筑中常常作为装饰的重点，被做透雕或彩绘处理。挂落主要有步步锦、灯笼框、冰裂纹、葵纹等纹样，还有很多其他的吉祥图案，如琴棋书画、雅宅众宝纹、八骏图、双凤朝阳等，寄托吉祥、平安等美好愿望。

ice cracks, sunflowers and so forth. There are also such auspicious patterns as images of zither, chess, calligraphy and painting, patterns of elegant house with all the treasures, pictures of eight spirited horses, two phoenixes darting toward the sun, and so on, bearing such good wishes as good fortune and peace.

> 门帘架

门帘架常位于明间（建筑正中的一间）隔扇的外面，用来挂门帘，也有次间用门帘架。除了有挂门帘的实用作用，门帘架还有很强的装饰性，门帘架上常常雕刻有各种图案。

> The Door Curtain Frame

The door curtain frame is often installed outside of the partition door of the room in the very center of the building to hang the door curtain. Sometimes it may also be seen in the side rooms. In addition to its practical function of hanging a curtain, the door curtain frame is also highly decorative and is often carved into a variety of patterns.

• 山西王家大院门帘架

门帘架上的三阳开泰图案周围点缀有花草，使图案更加丰富。刻有羊下巴上的胡子，使羊的形象更加生动；正中刻有太阳图案、松树，喻示长寿。

Door Curtain Frame at the Wang's Courtyard in Shanxi Province

The pattern of Three *Yang* Bringing out the Bliss on the door curtain frame is surrounded by floral decorations, making the overall pattern much more enriched. The beard on the chin of the goats (homonym to "*Yang*" in pronunciation) makes the goat images more vivid. In the center is a pattern of the sun. The pine trees symbolize longevity.

● 山西王家大院门帘架上雕刻的十鹿图

十头鹿的形象被一一展现在有限的画面里，鹿奔跑、跳跃的姿态被刻画得十分逼真，对十头鹿不同方位的处理，使画面显得错落有致、疏密适宜。

10-deer Pattern Carved into the Door Curtain Frame at the Wang's Courtyard in Shanxi Province

A picture of 10 deers is portrayed in the framework, with each one of them realistically depicted in different postures of running and galloping. The handling of the directions of the 10 deers makes the picture look well-proportioned in the space with appropriate density.

> 罩

罩主要被安装在室内柱间，用来分隔空间，集木雕、书法、绘画艺术等于一身，有很强的装饰功能。常见的有几腿罩、落地罩、栏杆罩、炕罩、圆光罩、八角罩等。罩多以硬木制作，雕工精细，纹饰美观。

> The Mask

The mask is mainly installed between indoor columns to separate the space. It comprises such elements as wood carving, calligraphy, artistic painting, and so on, and boasts strong decorative functions. Common masks include table-leg mask, down-to-ground mask, railing mask, the brick-bed mask, halo mask, octagonal mask, and so forth. Masks are mostly made of hardwood with fine carvings and splendid decorations.

太极殿炕罩

炕罩又叫“床罩”，是被专门安放在床榻前面，用来遮挡床榻的，内侧可以悬挂幔帐和软帘。其形式与落地罩相同。室内顶棚过高时，可在炕罩上加顶盖，在四周做毗卢帽、如意头等装饰。

Brick-bed Mask at Hall of the Supreme Principle

The brick-bed mask, also known as the bed mask, is specifically placed in front of the bed to mask the bed crouch. Curtains or soft drapes can be hung on the inside of it. It is in the same shape as the down-to-ground mask. If the ceiling of the room is too high, a cover may be added over the brick-bed mask, with monk-crown patterns, *Ruyi* head patterns or other ornaments on the four sides.

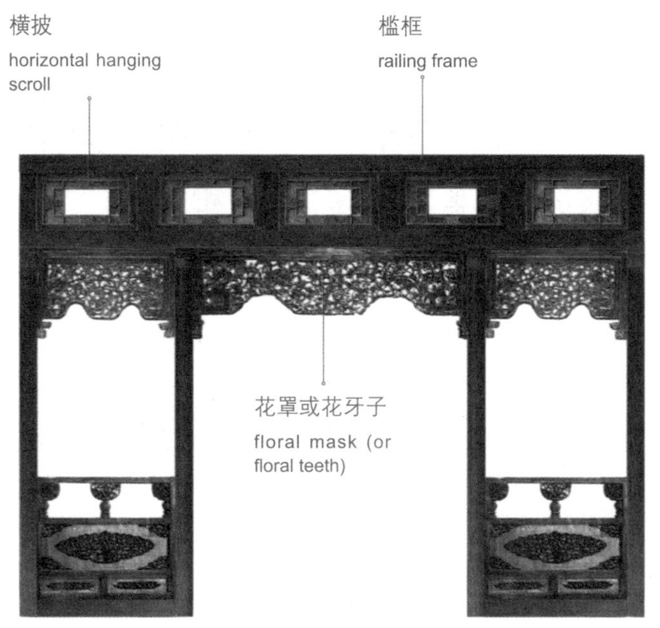

- **长春宫栏杆罩**

栏杆罩由槛框、横披、花罩、栏杆等组成。栏杆罩有四根框落地，两根立框，两根抱框，形成中间为主、两边为次的三开间形式。其中间与几腿罩相同，但两边下安栏杆，故名。这种罩多用于进深较大的房间。

Railing Mask at the Palace of Eternal Spring

The railing mask is made up of the railing frame, the horizontal hanging scroll, the floral mask and the railing. A railing mask has four down-to-ground posts, namely, two standing posts and two holding posts, shaping up a three-width form with the central part as the primary body and both sides as the secondary. The central part looks like a table-leg mask with railings at the bottom of both sides, hence the name "railing mask". Railing masks are mostly used in rooms with greater depth.

- **安徽宏村承志堂几腿罩**

整组罩有两根腿，腿与上槛、挂空槛组成几案式框架，两根抱框恰似几腿，故名，适用于进深不大的房间。

Table-leg Mask at Chengzhi Hall in Hong Village, Anhui Province

The whole set of mask has two legs. Together with the upper rail and the hanging rail, they shape up a frame like one side of a tea table, hence the name "table-leg mask". It suits rooms of smaller depth.

• 北京北海公园静心斋落地罩

落地花罩形式与几腿罩相近，不同之处是横批下面的花罩沿两侧抱框向下延伸，到达地面，接须弥墩。这种罩的面积较大，装饰性很强。

Down-to-ground Mask at the Jingxin Study in Beihai Park, Beijing

The down-to-ground mask looks very similar to table-leg mask in terms of pattern. The difference lies in the floral mask at the bottom of the horizontal hanging scroll, which crawls along the holding posts on both sides all the way down to the floor at the connection with the Buddhist Sumeru base. The down-to-ground mask is quite big and offers great decorative scope.

• 苏州耦园还研斋圆光罩

圆光罩、八角罩被安装在室内进深柱间，做满装饰，中间留出圆形、八角形或其他形状的门洞，门洞内可以悬挂软帘，用来划分室内空间。

Halo Mask at the Huanyan Study in the Lotus Root Garden, Suzhou

The halo mask and octagonal mask are installed among pillars in rooms with depth to provide full-coverage decoration with a hole in the shape of a circle, an octagon or others, from which soft curtains can be hung for the purpose of separating the space inside the room.

> 天花

天花又称“顶棚”，是室内顶部的装饰形式，还可以保暖、防尘。天花一般分为硬天花和软天花。硬天花以木条纵横相交成若干格，每格上覆盖木板，称“天花板”，天花板圆光中心常绘有龙、

> The Ceiling Pattern

The ceiling pattern, or the plafond, refers to the decoration of the roof indoors. It can keep in the warmth, and protect against dust. Generally, there are two types of ceilings, the hard one and the soft one. The hard ceiling is usually structured into grids by wood strips

- **团龙平棋**

由木肋格子和装填木板构成的大方格犹如棋盘，因此被称为“平棋天花”。格子上有彩绘或者雕花作为装饰。

Loong Chessboard

The big square pattern composed of wood grids and covering boards bears a resemblance to a chessboard, which gives it the title of "chessboard ceiling". The grids are decorated with colored paintings or carvings.

- **海墁天花**

海墁天花表面平坦，没有肋条，以木板做成，或是在较小的房间内架起一个完整的框架，上面安装木制平板或糊纸，顶部彩绘为一整片的形式。

Haiman Ceiling Patterns

The surface of the *Haiman* ceiling patterns is flat with no ribs. It is made of a wood board or a complete piece of frame, in the case of a smaller room, installed with a flat wood panel or pasted paper on it with the whole surface colorfully painted.

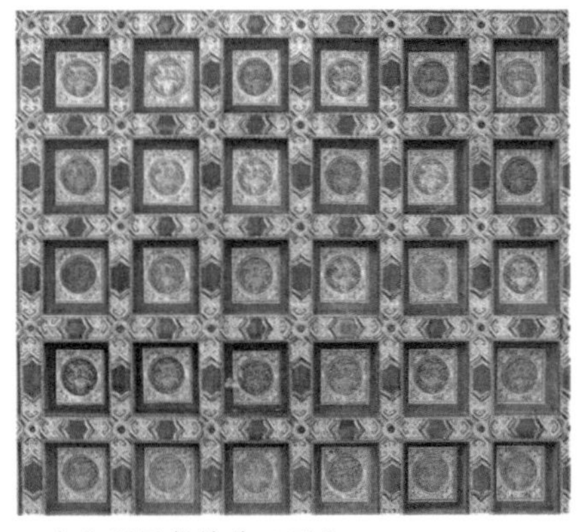

- **太和门团龙纹井口天花**

由“井”字形木肋格子组成的天花，是天花的最高形制，清代时多用于宫殿建筑。

Compartment Ceiling with Coiled Loong Patterns at the Gate of the Supreme Harmony

The ceiling pattern is composed of grids made of wood ribs in the shape of the Chinese character "井 (well)". It is the highest form of ceiling patterns, mostly used in imperial palaces in the Qing Dynasty (1616-1911).

凤、吉祥花卉等图案。软天花又称“海墁天花”，以木格篦为骨架，满糊麻布和纸，上绘彩画或用编织物装饰。

vertically and horizontally intersected with one another. Each grid is covered with a board called ceiling board. The central halo of the ceiling board is usually painted with patterns of loongs, phoenixes, auspicious flowers, etc. The soft ceiling, also called "*Haiman* ceiling", is made of wood castor-oil plant as the frame, pasted with sackcloth and paper with painted patterns or basketwork.

> 藻井

藻井是传统建筑室内顶棚的独特装饰部分。“藻”即水藻，代表水；“井”为天文上所称的“东井”，为贮水之所。藻井是传统建筑等级的一种体现，多用于宫殿宝座和寺庙佛龛上的屋顶，南方的祠堂、戏台也使用藻井。藻井呈穹隆

> The Caisson

The caisson (*Zaojing* in Chinese) is the unique decorative part of the indoor ceiling in traditional architecture. "*Zao*" means algae, which stands for water; "*Jing*" refers to the Chinese astronomic terms of "east well", a place to store water, and embodies the ranks in traditional architectures. Caisson is used

• **四方形藻井**

四方形藻井又称“斗四藻井”，表面呈四方形，较为常见。

Square Caisson

Square caisson, also know as "*Dousi* Caisson", is mostly seen with squares on the surface.

• **圆形藻井**

福建泉州天后宫进门处戏台的圆形藻井，穹隆状，似斗。

Round Caisson

Round caisson of the theater stage at the entrance of the Palace of the Queen of Heaven in Quanzhou, Fujian Province, takes the form of a dome like a funnel.

• 北京故宫澄瑞亭的八角藻井
Octagonal Caisson of the Pavilion of Auspicious Clarity in the Forbidden City, Beijing

状，大多以斗拱层层相叠而成，简单的藻井则以木板制作。藻井的形制有四方形、圆形、八角形等，也有将几种形状融为一体的，精美而又华丽。

in the dome over the throne in the palace or over the shrine in the temple. In south China, it is also used in ancestral shrines or theaters. A caisson takes the shape of a dome. It is based on wood brackets put together layer by layer. The simple version of caisson used wood boards in its construction. There are square caissons, round ones, octagonal ones and so on. In some cases, a combination of several different types can be seen of great delicacy and splendor.

太和殿的盘龙藻井

北京故宫太和殿金銮宝座正上方的天花板上有一个盘龙藻井。此藻井分上、中、下三层，上圆下方，以斗拱承载，通高1.8米，井口直径达6米，顶部中央为圆形盖板，周围施以28攒组成的小斗拱。穹隆圆顶心的明镜下盘卧着一条巨龙，口衔宝珠，俯首下视。龙口所衔宝珠名“轩辕镜”，为青铜制成，外涂水银，光亮如镜。在宝座的正上方悬轩辕镜，喻示皇帝得位于大统，同时可保大殿安全。

Caisson with Coiled Loongs in the Hall of the Supreme Harmony

Over the golden throne in the Hall of the Supreme Harmony inside the Forbidden City in Beijing is a caisson with patterns of coiled loongs. It has upper, middle and lower layers, with the upper layer being round and the lower one square. Each layer is supported by brackets and tapers to the top with a vertical height of 1.8 meters and a diameter of as wide as 6 meters. In the center of the top is the round cover board, surrounded by 28 sets of small brackets. Under the clear mirror in the middle of the top dome lies a giant coiled loong with a pearl in its mouth, looking downward. The pearl, called "*Xuanyuan* mirror", is made of bronze, coated with mercury, making it as bright as a mirror. It is hung directly over the throne to indicate that the emperor is enthroned by the great unity. It also implies that the hall would be safeguarded .

太和殿盘龙藻井

顶部绘有或雕刻有盘龙纹的藻井是最高等级的藻井，仅用于与皇帝有关的宫殿和最高等级的礼制坛庙。

Caisson with Coiled Loongs in the Hall of the Supreme Harmony

The top of the caisson is painted or carved with the pattern of coiled loongs. It is the highest-class caisson, used only in the emperor's palaces or the temples or shrines of the highest level in accordance with the ritual system.

> 碧纱橱

碧纱橱又作“壁纱橱”，是一种做工精细、通透的隔扇，由槛框、横披、隔扇等部分组成，被安装在室内进深的柱间。根据房间的大小，碧纱橱由6扇至12扇的隔扇组成，其中只有两扇是活动扇，作为进出口。碧纱橱用料讲究，其上有精细的雕刻，上层的仔屉内有各种形式的棂条花纹。仔屉采用夹堂做法，屉有两层，中间被夹入纱或玻璃，上面绘有花草鱼虫、人物故事，或题有诗词和格言，具有很高的观赏价值。

> The Green Gauze Cabinet

The green gauze cabinet, or "wall gauze cabinet", is a fine work of transparent partition door made up of the frame, the horizontal hanging scroll and the partition door, installed between pillars in a room with ample depth. By size of the room, the green gauze cabinet may be composed of 6 to 12 partition doors, only two of which are movable to gain entry. The green gauze cabinet lays much stress on the selection of the materials, and intricate carvings are made on it. In the lattice on the upper part of it are various forms of lattice strips and floral patterns. The lattice is built of clipped frames in two layers with gauze or glass inserted in between. Patterns of flowers, grass, fish and insects, story pictures or inscriptions of poems or maxims are painted on it, endowing it with a high value of appreciation.

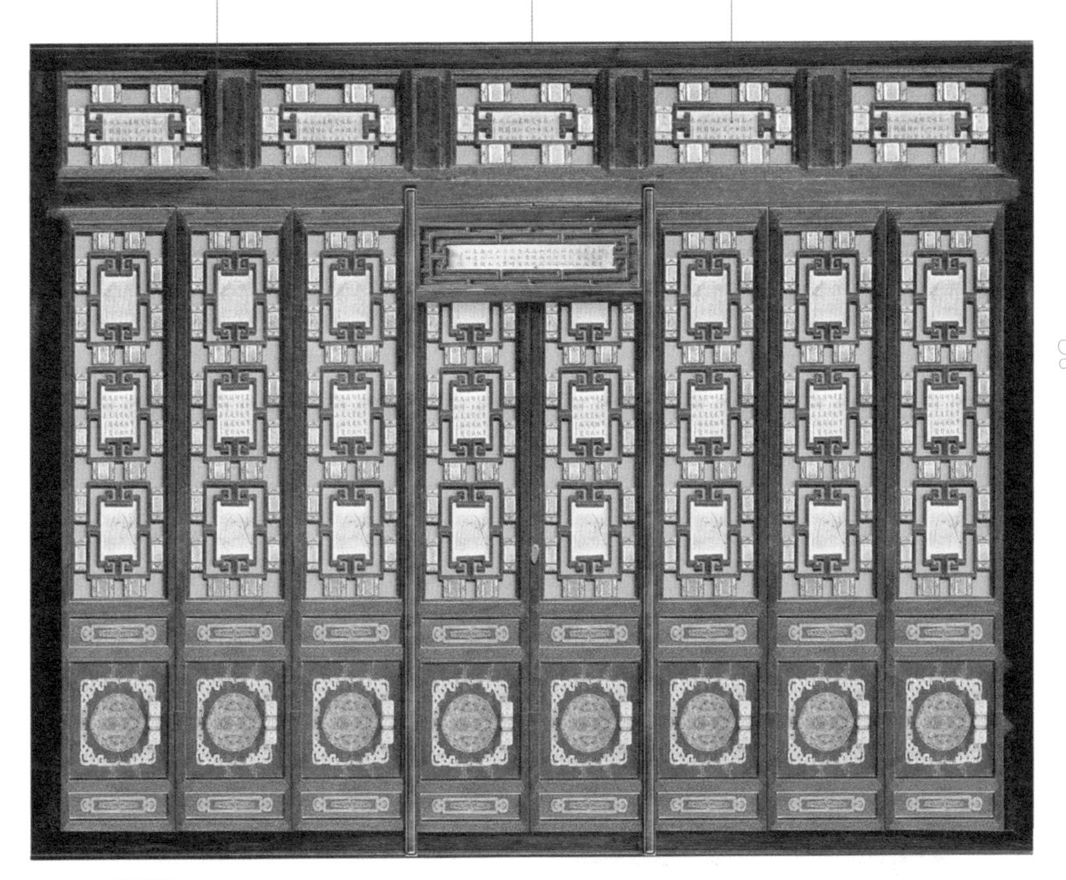

• 碧纱橱

Green Gauze Cabinet

> 太师壁

太师壁指明堂后檐金柱间的壁面装饰，多见于南方园林厅堂之中。壁前被放置条几案等家具及各种陈设。太师壁起于地面，上至屋梁，左、右两侧留有空间供人通行。壁面有多种样式，有的以雕刻为主，有的由隔扇组合而成，有的采用棂条槛窗的形式，有的则在板壁上刻字、挂画。

> The *Taishi* Wall

The *Taishi* wall is a wall decoration installed between the hypostyle columns of the back eave behind the main hall. It is often seen in the halls in gardens in south China. In front of it is put such furniture as long tables and various furnishings. The *Taishi* wall starts from the floor and rises up to the beam of the house with reserved space on its both sides to allow passage. The surface of the wall is variously featured, sometimes with carvings and other times with combined partition doors or grid-patterned windows. It may also take the shape of either the window frame with lattice bars or a board wall with carved texts or hung scrolls.

• 安徽程大位故居维新堂太师壁
Taishi Wall at Weixin Hall of Cheng Dawei's Former Residence, Anhui Province

> 博古架

博古架是用来陈设各种古玩的木架子。明清时期，人们在装饰室内时，在柱间设置不可移动的博古架，以隔断室内空间并做陈设家具之用。一般分为上、下两段。上段为博古架，被分出很多形状不同的空格，可在空格中陈列各种古玩，极其雅致。下段是有柜门的小柜橱，用来储藏古玩。如果由博古架分隔开的两个空间需要联通，还可在博古架的中间或两侧留出通道，供人通行。博古架花格通透，被陈列在其中的古玩显得富贵而典雅。

> The Antique Shelf

The antique shelf (*Bogu Jia* in Chinese) is a wood shelf used to display various antiques. In the Ming and Qing dynasties (1368-1911), the antique shelf was usually set immovably between pillars as interior decoration in order to separate the interior space and display articles. The shelf can be divided into two sections: the upper section is the antique shelf with spaces of different shapes to display various antiques in a very elegant way; the lower section is the small cabinet with doors to store antique articles. Sometimes, a passage may be spared in the middle or either side of the antique shelf to allow people to pass through between the spaces it divides. The open-space shelf serves well in striking a rich and elegant note in its exhibition of antiques.

博古架上部栩栩如生的花鸟图案

vivid patterns of flowers and birds shown in the upper part of the antique shelf

博古架顶部的龙形木雕

woodcarving in the shape of a loong at the top of the antique shelf

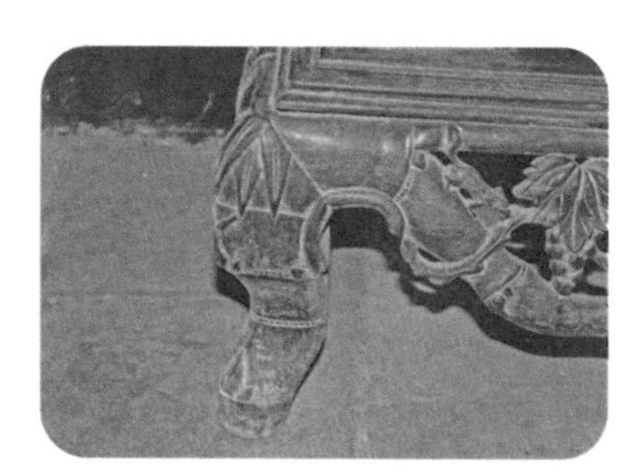

博古架腿上精致的植物纹样

exquisite patterns of vegetation shown on the foot of the antique shelf

• 博古架

Antique Shelf

博古

所有的吉祥器物统称“博古”。建筑上雕刻的暗八仙、如意、贡果、花瓶、琴棋书画等具有吉祥寓意的纹饰都可以叫作“博古纹”，它是北宋以后中国古代常用的装饰性纹样。以这类题材作为装饰，有高洁、清雅之寓意。

Bogu

All auspicious objects are collectively referred to as *Bogu*. In architecture, any carvings with such auspicious items as an implied eight immortals pattern, *Ruyi*, tribute fruit, vases, and the combination of zither, chess, calligraphy and painting, and so on are called the pattern of *Bogu*. The pattern of *Bogu* has been massively used as decorative patterns since the Northern Song Dynasty (960-1127). As a decorative theme, the pattern imparts a sense of nobility and elegance.

• 家具上的博古纹
Bogu of Furniture